Simple Solutions to Energy Calculations

Simple Solutions to Energy Calculations

Richard Vaillencourt

Published by
THE FAIRMONT PRESS, INC.
700 Indian Trail
Lilburn, GA 30247

Library of Congress Cataloging-in-Publication Data

Vaillencourt, Richard, 1951- .
Simple solutions to energy calculations / by Richard Vaillencourt.
p. cm.
Includes index.
ISBN 0-88173-181-1
1. Energy conservation. 2. Energy auditing. I. Title.
TJ163.3.V35 1994 696-dc20 94-8311
CIP

Simple Solutions To Energy Calculations by Richard Vaillencourt.

Published by The Fairmont Press, Inc.
700 Indian Trail
Lilburn, GA 30247

Printed in the United States of America

10 9 8 7 6 5 4 3 2 1

ISBN 0-88173-181-1 FP

ISBN 0-13-148388-9 PH

Distributed by PTR Prentice Hall
Prentice-Hall, Inc.
A Paramount Communications Company
Englewood Cliffs, NJ 07632

Prentice-Hall International (UK) Limited, London
Prentice-Hall of Australia Pty. Limited, Sydney
Prentice-Hall Canada Inc., Toronto
Prentice-Hall Hispanoamericana, S.A., Mexico
Prentice-Hall of India Private Limited, New Delhi
Prentice-Hall of Japan, Inc., Tokyo
Simon & Schuster Asia Pte. Ltd., Singapore
Editora Prentice-Hall do Brasil, Ltda., Rio de Janeiro

Table of Contents

Introduction

Most energy conservation projects are implemented only after passing through several levels of analysis and decision making. It is obvious that it is not cost effective to spend a lot of time on the initial levels in this process. That does not mean that the information gathered and methods of analysis do not require accuracy and technical correctness. On the contrary, there is no time that it is acceptable to use inaccurate or technically flawed data or methods. On the other hand, it is foolhardy to waste too much time trying to nail things down to the last decimal place when an analysis based on the information that you already have would indicate that the measure is economically unacceptable, or so good that it's a sure thing.

Since I must admit that I am undeniably slow and lazy, I wanted some tools to quickly and easily answer the question: "Is this project worth spending time and money to get detailed answers?" I sure didn't want to pay someone to tell me that my idea was ridiculous. Worse yet, I didn't want my boss to think that I couldn't figure it out myself.

Some energy evaluations are quite easy. Some others are extremely difficult, but mostly unnecessary at this stage. The engineers who write the energy auditing handbooks seldom take the trouble to show you how easy the easy evaluations are. They seem to desire to impress you with their understanding of the difficult ones by going into great detail whether you want to or not.

To solve my dilemma, I developed several "plug and chug" computer spreadsheets for various energy evaluations that seemed to occur over and over. Lighting, occupancy sensors, variable speed drives for fans and pumps, etc., are some of the areas that are common to any facility. My simple solutions are presented here, hopefully, to make your life easier. Perhaps you need to do this type of analysis often enough that you can develop your own spreadsheet specific to your needs. I offer them to all the other slow and lazy engineers out there so that you can stay under cover.

I also admit to having a high degree of anxiety about going public with my methods. I am sure that there are many very smart engineers who will attack this book on the grounds that it commits several serious errors.

First, its goal is to answer only one question. Namely, "Should I drop this project or look deeper?" If you asked them (at $200/hour) they would feel obligated to deliver several pages of calculations to justify their fee.

Second, they will complain that I made almost no attempt to justify, derive, or demonstrate with rigid completeness the validity of my methods. I told you I was lazy. If you use them and they seem to work and you want to know why – go for it! The proof is left to the student.

Finally, they will say that the solutions are simple because I left the complicated things out. You bet I did! I told you I was slow. That is why I wrote this book. I believe that you can grasp the significance of a project's worth and outline its limits and feasibility without being forced to work through the intricate, unnecessary, and mostly irrelevant engineering gymnastics so faithfully performed by unpractical engineers. Proficiency at doing the unnecessary does not increase efficiency.

The only warning is to remind you, again, that the purpose of this approach is to provide a decision making tool to help you to decide whether to invest real time and money into developing the details of a project. Don't bet your job on the exact value of the savings calculated from these equations. Your recommendation should be: 1) no, this won't pay; or, 2) yes, this looks good so far and should be the target of an in depth study.

Chapter 1

The Walk Through Audit

INTRODUCTION

The saying goes that a journey of a thousand miles starts with the first step. The same applies to the journey to a completed Energy Conservation Measure (ECM). The first step is to have an idea.

There are many ways to get any particular ECM idea. You can read about someone else doing it. Or you can have a vendor (who is trying to sell something) point it out to you. Or you can have your wife (spouse) say, "Hey! Look, stupid! If you put a variable speed drive on your cooling tower fan you can save a lot of money." Most of the time your ideas will come from walking through the facility with your eyes and ears and mind open.

This is called a "Walk Through Audit". Why is it called an audit? Why not a "Survey" or "Analysis" or some other engineering term? Because the federal government created this term. Spoken like true accountants, they were hoping to create more accountants out of engineers. The original thrust of the activity that they were trying to define (and therefore regulate) involved balancing (auditing) the energy books. As accountants, they feel that it is important to identify the energy (cash) flow. They can't help it. Their need to balance the books is either implanted into their psyche in college or is the genetic defect that caused them to become accountants. Whatever the cause, the demand for a one page, all encompassing view is in sharp contrast to the engineering approach of solving a large problem by solving all its smaller parts first.

There is an argument that says that without the "whole picture" you will run the danger of wasting your opportunities. This is absurd. If the evaluation indicates that you will save 10,000 Btu or kWh with a two year payback, that is real energy and real savings. It does not matter if the 10,000 units were a small part of a major consumer or the entire consumption of a smaller consumer. The dollars saved and the dollars spent are absolute numbers. If it is reliable and cost effective, do it. Stop wasting time trying to find "The Perfect Project". Think of the money you saved by not spending the time on the overview.

Anyway, using the accounting term makes it easier to get them to read the report. Since the accountants control the money it's better to humor them.

Most walk through audits are not formal affairs. The really good plant engineers are doing it when they walk through the plant to do their normal work. They are always looking for a better way to get something done. Not just energy, but: storage, material handling, purchasing supplies, water, sewage, pollution control, hazardous waste management, maintenance, installations, demolitions, etc., etc., etc. This informal, but very effective walk through audit can produce some very significant results because a good plant engineer knows where the skeletons are buried.

However, bear in mind that this productivity can be improved by a formal walk through audit. If you take the time to concentrate only on energy utilization and to record your observations you will start to see the individual trees in the forest in front of you. Be careful. Sometimes the forest turns out to be a murky swamp.

This productivity can be amplified if the walk through audit is performed by an experienced energy professional. Why? Because the intimate knowledge of the facility in the brain of the plant engineer can be tapped by another engineer and combined with his specialized training and experience to develop a larger list of potential projects than either engineer can do by himself.

A walk through energy audit consists of two distinct operations. The first is to determine with as much accuracy as is possible exactly what the existing conditions are. This is accomplished during the walk-through stage. This is most difficult and most important. If you incorrectly establish that a certain piece of equipment utilizes several times more energy than it really uses, you will just as incorrectly calculate a larger savings than is possible. The extreme case (that I have seen too often in reports) is that you will calculate energy savings that exceed the annual consumption of the entire facility. The second operation is defining and evaluating (ECM). That will be addressed later in this book.

The walk through audit report is meant to be a first cut report only. The idea is to identify measures that appear to be worth a detailed evaluation. Many of these measures will not pass a second cut. Quite often due to feasibility or product quality conflicts, not economics. Your job is to point out the opportunity, ballpark the potential, and make them tell you "why not".

You have accomplished your goal. The measure has not been overlooked. It has been evaluated and discarded for (hopefully) logical reasons.

The limitations of a walk-through audit should be clearly understood. You cannot completely understand any complicated processes and even most simple ones as they apply in any specific application. However, the process auxiliaries will, in most cases, be standard applications of pumps, fans, cooling (air and water), lighting, etc. The goal is to suggest ways to provide these functions in such a way that the process is "unaware" of a change, yet the operating expense is reduced and reliability is unchanged, or improved.

This approach will not produce detailed plans and specifications. The goal is to point out the possibilities with enough detail to get a feel for the difficulties, and cost and savings estimates reasonable enough to indicate which ideas deserve an in depth study.

The best and most often used tool you can have is common sense. You don't need a college degree, but a certificate from the school of hard knocks will help. You must always ask yourself two questions: 1) Will there really be a savings? and, 2) Will many people complain about the change?

A dictionary definition of the word "engineering" is: "1. the science of making practical application of pure sciences." Similarly, the word "art" has, as two of its meanings: "1. the principles governing any craft or skill. 2. skilled workmanship or execution." We, as energy engineers, must strive to employ the practical application of our training and experience, with a major emphasis on being practical.

Why? Because in the world of building managers and production managers energy conservation has been a synonym for suffering. As energy engineers we really haven't done much to dispel these fears. In many cases we have earned the reputation for going off half-cocked. A reputation for squeezing off the flow of Btus with little regard for anything else.

To counter this we must continuously learn the "principles governing our craft". And since many of these principles affect non-engineering facts of life, we must apply them skillfully. We must know when they are important enough to push for and when to back off. This is art. The elements of style, finesse, subtlety, etc., must be constantly balanced with facts and reality. Engineering realities, financial realities, cultural realities, and the most frustrating of all: human nature.

Have you ever wondered why a production supervisor is not pleased to see you? And if not openly hostile, he is at best uncooperative.

That's because Energy Management means changing his procedures, retraining his men, and giving him more paperwork. He is a successful production supervisor because he produces. Why should he go out of his way to do something that will only make you look good? And give him more headaches. What's in it for him?

It's all right to tell him about making the company more competitive or profitable, and I'm sure he agrees with those goals. But my experience has shown me that in reality his next raise or promotion depends entirely upon whether he gets the product out on time without mistakes. What does that have to do with energy management? Nothing. That's your problem, not his. Your next raise or promotion depends on how much you reduce the company's energy bills. Nobody comes to you and asks you why orders are not being filled on time.

> *"We should not expect to utilize in practice all the motive power of combustibles. The attempts made to attain this result would be far more hurtful than useful if they caused other important considerations to be neglected. The economy of the combustible is only one of the conditions to be fulfilled in heat engines. In many cases it is only secondary. It should often give precedence to safety, to strength, to the durability of the engine, to the small space which it must occupy, to small cost of installation, etc."*
>
> Sadi Carnot; 1824

To put it in another perspective, let's look at the two ways of making a baby. By far, the most efficient and reliable method is to mix egg and sperm in a test tube then implant the resulting gamete in the "appropriate environment" for development. This is much more reliable than the random biological chance of the egg and sperm meeting while coincidentally already in the "appropriate environment" at the same time. It certainly uses a lot less energy on everybody's part. And, if you throw in the cost of several dates, and engagement ring, the wedding ring, the wedding, the reception and the honeymoon (I realize that this is the old fashioned approach), the test tube method is much more cost effective.

Personally, I refuse to give up the biological method. As a matter of fact, I have been known to expend all that time and energy in a biological attempt that I sincerely hoped would fail! So the first question: "Would there be any real savings?" is answered with a resounding, "NO!". Laboratories can rebate, government can mandate and give tax credits, and I still will want to use the biological approach. Salesmen can offer me trips to a tropical paradise as an incentive, and that will only confuse me because I would want to go there only in the hope of increasing my biological attempts. So the answer to the second question, "Will many people complain about the change?" is affirmative, and usually unprintable.

What does this have to do with a walk through audit? Everything. To get the plant engineer on your side you must state clearly, and more importantly, you must *believe* that efficiency is not the only parameter that merits your attention. You only have a few hours to spend with the plant engineer. You need him to keep you out of trouble. You need him on your side. You need to convince him that you are not blind to his needs and are not going to make a name for yourself without any care for his problems.

By far, the hardest thing to overcome when dealing with a client plant engineer is that he is, in many cases, defensive. This is completely understandable.

First: He lives, breathes, eats and sleeps with his facility. Whether he wants them to, or not, the plant personnel will find him when they need him.

Second: As I mentioned above, he is always looking for a better way to get something done. Most of the time some "bean counter" is pushing an idea that he read about in Popular Science. Having been there, I know how you can start with a chip on your shoulder. All it takes is for someone to come in off the street, make a proposal to some production manager (that is the same thing you have been talking about for two years) that gets approved in two weeks, and you can get bitter.

Third: There is no way that anyone off the street can know his plant and processes better than him.

Fourth: He has the attitude: If there was a better way to do it, I would have done it already.

These attitudes are very difficult to avoid developing if you are the plant engineer, and very difficult to overcome if you are the one coming in "off the street".

As a conservation engineer, remember that it is his plant and he has dealt with a lot of "snake oil salesmen". They waste his time and annoy the daylights out of him. Admit immediately that you know he knows more than you about his plant. But try and point out that he shouldn't be expected to know everything about the latest technology. Maybe you do have something to offer. Together, you might be able to sort out what will, or will not, work for him.

The hardest thing for an energy auditor to deal with is the wrong information. There is no way (see #3, above) that you can tell that it's wrong. If you're smart, you will take a long time to describe what you think will happen, or needs to happen if a certain ECM is going to work. Hopefully, in this discussion, if it is a real dialogue, enough information will be passed between the two engineers so that both the auditor and the plant engineer know enough to keep each other out of trouble.

That is the essence of this last discussion. KEEP EACH OTHER OUT OF TROUBLE.

If this attitude is made clear to the customer you will be seen as an asset. Your goals are the same as his. You are there to present options. To improve the profitability and the productivity. You are really there to help him.

If you're a vendor and you feel that the client really doesn't know the answers to important questions, don't stop asking and hope everything will turn out right. If it's important enough, an obvious wrong answer might lead you to the conclusion that the best course of action is to walk away from the project.

Give the plant engineer full power of veto! You may think that you know what will work but the plant engineer knows what will be *accepted.* If he indicates that a certain measure is not worth it to him, by all means try to find out why and even try to change his mind. But if he still insists – drop it!

I can't say this strong enough! You don't have to know or accept his reasons. You don't have to agree with him. You do have to give him that power. It may be that he has been trying to get funding for the exact same ECM for the last five years and is just plain tired of fighting (and losing) over it. Or, it may be that he thinks that it is absolutely brilliant. So brilliant that he wants you to leave it out so that he can suggest it after you're gone. It doesn't matter. If he wants it left out you won't be able to get around him anyway so why make an enemy? If you push the plant engineer unnecessarily I guarantee the project is doomed to failure. If you prove to him that you are on his side and he has veto power, you will get the first projects implemented. Now you can build from this success to push for the ones you held back before, if you still want to.

It is my recommendation that you purchase, develop, copy, or steal an audit checklist form. You only have a short time to perform the audit and usually no access for a second visit to get what you missed. Trust me. You will not get everything. But with a good form to prompt you, you won't forget the obvious things.

Find a form and always be ready to modify it. It doesn't matter how many pages you carry. Obviously many of the pages will not apply to the plant that you are doing. If you have sections designed for a sewage treatment plant you won't use them in a grocery store. But the fact that you note "N/A" where appropriate under the "gas fuel" section will give you the confidence that you did not forget it entirely.

The last page should be a list of questions left unanswered when you left, but promised to be answered by plant personnel. A copy of this page with a delivery date should be left with the person at the plant responsible for getting the answers to you.

Keep this list short! Usually this consists of energy consumption records and costs, specific equipment operating times and strategies, horsepower of inaccessible motors, etc. Try very hard to keep it to information that you would like to have. And also try to limit the questions to ones that the plant engineer can delegate and nag someone else about. If he has to bug the accounting department for the fuel and electric bills he will enjoy it so much that you can be sure it will happen. It is rare for him to have the accountants owe him some information. The momentary feeling of power is intoxicating.

If there is a piece of information that you feel that you need to do your work, try very hard to get it while you are there. Many times the best intentions of the plant personnel are forgotten when they are faced with putting out daily fires. If the information is necessary for a complete and accurate report, you may be "on hold" for a long time. It is best to be ready to make an assumption (spelled "g-u-e-s-s") if the information is not forthcoming. At this stage assumptions are acceptable if they are clearly identified in the report as your best guess. It is much better to produce a list of ECM with a potential, but known flaw in it, than to wait indefinitely for details.

I suggest that you avoid making recommendations for production processes unless you are specifically asked. It doesn't matter that you were the Corporate Energy Manager for a textile company for over five years. All that means is that you can easily recognize textile machinery and its basic function. I know that this is a difficult restraint because the really big savings are in production conservation measures, not to mention the most fun. But you have to walk before you can run. Do a really good job on the easy stuff and you might get the chance to have some real fun. Screw up on a lighting measure and there will be a shoot-on-site order put out on you.

I usually ask if there is any "pet project" that they would like my help in getting evaluated and approved. After about two years as a Corporate Energy Manager I had pretty much accomplished the easy stuff. The "no brainer" projects with major savings and almost no capital. Now I wanted to tackle some of the more interesting ideas that were a little more risky and certainly more expensive. At the very least, they required a lot of thinking. It was very frustrating to send report after report to what I came to call "the Black Hole": my boss' desk.

Nothing ever came back from there. Not even acknowledgment that he received them.

It got to the point where I was asking a vendor that I trusted to take my evaluations and recommendations and submit them on *his* letterhead. Quite often that worked. If the Outside Consultant Syndrome exists at your plant, you might as well use it!

So, as an Outside Consultant, I offer to evaluate the energy part of some in house pet project. It is important to emphasize that the production part is *still* the responsibility of the plant personnel. Your role is to determine how much energy, or energy dollars, can be saved *if* this project will work without affecting production or quality.

What you should normally look at is all the auxiliary equipment and functions that support the process. Specifically, cooling water, heating method, exhaust, fume/dust collection, etc. The goal is to provide all these services for the *least cost* such that the process never experiences a change in its ability to function.

I used the term "least cost" on purpose. In order to be useful to the plant, your goal must be to save money. Certainly you are concentrating on the energy consumption. But the reason you are looking at the energy consumption is to reduce its cost to the operation as a whole. If energy didn't cost anything, you wouldn't be reading this book. There would be no need to conserve. As soon as it became expensive, the term Energy Manager was invented.

My own experiences with failed conservation measures taught me an important lesson: Most of the failures occurred when I went for every last Btu. That's when more than one person noticed that I changed something. That meant that they could talk to each other and complain as a group. Actually, I have to admit that I *have* gone just a bit too far once or twice. That's when performance fell into the "marginal" or "unsatisfactory" category (not enough Btuh in or out to meet the peak conditions, etc.). The surprise was that the entire project was considered a failure and discarded! I was never asked to just back off a little. It always went back to what was there before.

The goal of the previous discussion is to help an outside consultant develop a style and attitude that will encourage cooperation and support from inside the plant. Such support will improve the accuracy and the volume of information that you can acquire in the short time provided. In house engineers can use many of the elements on colleagues and plant personnel.

I know that it takes a lot of courage and personal restraint to go to a production supervisor and let him know that you are looking at changing something in his little kingdom. He will invariably start off with, "You want to do what?!" followed by either a loud laugh or a loud groan.

Strangling him will only give you a momentary pleasure. And it will look bad on your resume.

His second reaction will be, "You can't do that because (fill in the blank) ".

1. it'll never work
2. we tried that before you came and it didn't work

3. it's too complicated for the operators
4. it'll ruin quality
5. the tenants/customers won't like it
6. it'll ruin our sales
7. the boss/owner won't like it

Any one of the above is a valid reason to walk away from a project. The problem is that they are said without regard for the truth only to get you to walk away. They are thrown at you like hand grenades that you have to throw back (prove that they aren't true) before they explode. The good news is that, like a hand grenade, if they are thrown back fast enough they explode on the other guy. Once he has been blown up by his own grenade he has to cooperate with you or be identified as an obstructionist. Or worse in corporate circles, he'll be labeled: "Not A Team Player".

At all times, whenever possible, for as long as possible, talk to the front line people. The people who run the machines, open and close the valves, turn the lights on and off, set the thermostats, change the light bulbs, etc., etc. You will certainly run the risk of meeting the plant Bullwinkle (Mr. Know-It-All) and the plant Archie Bunker (everything's being done wrong). But the vast majority of these people know what's going on for real. They also know when something can be improved, and when it can't. They will be very happy to tell you exactly what's on their mind. Most of the time it's information that is worth taking the time to get.

As you perform your calculations, you start to walk a fine line between engineering and opinion. The final numbers are only as good as your assumptions. The best approach is to take the attitude: "This is what I thought I saw and what I thought I heard. If these assumptions are reasonably correct then the report is reasonably correct."

Making your assumptions is like playing "The Price Is Right". You should always be trying to get as close as possible to the actual facts without going over. The best assumptions are the ones that are honest and slightly conservative. The idea is to keep from overstating the possible savings while at the same time avoid grossly understating the savings to such a point that a promising opportunity is discarded.

If at all possible, get the plant engineer or the front line operator to make the assumptions for you. Most of the time they will have a better idea about reality as it exists in their world than you will. More importantly, they won't be able to accuse you of influencing the outcome in your favor. (I have never been able to figure out how I would benefit from suggesting an ECM, but I have been accused of somehow acting in my own self-interest.)

By all means try to keep them honest. If they are making assumptions that are outrageous gently bring them back to reality. I have seen attempts to justify an ECM by assuming the annual run time was 9,000 hours per year. There is no way to assume that there are more than 8,760 hours unless it's leap year or you are a lawyer billing a client.

Utility Costs

There is a lot of wasted effort put into determining the "correct" utility costs to use in a calculation.

Much of this is the result of confusing what is needed for energy management and what is needed to evaluate a retrofit project. The fine details of ON PEAK, OFF PEAK, SHOULDER, etc. rate structures are necessary to understand the effects of day to day operations. When you are an "in house" energy manager it should be your job to know these details and know them well. But to evaluate a retrofit project identified in a walk through audit all you need is an understanding of the seasonal rates and whether the project has a seasonal operating schedule. After that, all you need to determine is the facility's average incremental costs for demand and consumption.

I know this is heresy.

It always amazes me that the cost of oil and natural gas can fluctuate every two weeks and it is perfectly acceptable to use the annual average cost. It is even acceptable to use an annual weighted average when there is dual fuel capability that allows the facility to switch to the cheapest fuel several times a year. But when it comes to electricity suddenly it becomes necessary to do an hourly analysis. Not only is there this double standard, but what does the extra effort get you or the customer? If an evaluation is performed using the billing charges for peak demand and the average kWh costs, and done a second time paying careful attention to the various kWh adjustments and time of day variations, the differences will not be significant enough to mention. I have made this comparison and found the differences to be less than 5%. And who knows which one is right?

CALCULATIONS

The energy cost calculations need to be done before any measure can be evaluated. These are usually simple averages. If you are contemplating a measure that truly is used seasonally, such as most air conditioning at a school, it will become necessary to calculate the seasonal energy costs. The method is the same, you just have to perform the calculations for discrete periods of the year.

In all cases, you start with at least twelve consecutive months of energy bills.

Incremental Electrical Demand (kW) Charges

Most rates today are not declining block rates. Therefore, the incremental demand charge is the charge shown on the bill divided by the peak demand shown on the bill. Otherwise, if you have a copy of the utility rate structure, the demand charge will be clearly shown there. But beware, rates change very often these days. If you don't have an up to date copy of the rate you will probably be calculating lower costs.

Most utilities only charge for the peak demand that occurs during their peak usage period. If you are evaluating a measure that will reduce the demand during their off peak usage period, i.e., nighttime parking lot lights, it will not reduce the customer's demand charges. The customer probably isn't paying any demand charges for the use of those fixtures. Always be careful that you are not claiming a reduction in a charge that doesn't exist. Some people might think that you don't know what you are doing. Others that believed your numbers will be disappointed when the savings don't show up.

Incremental Electrical Energy (kWh) Charges

Once you have determined the proper demand charge and the demand billed for a certain month, it is easy to determine the incremental cost per kWh. Simply subtract the demand charge from the total charges. The remaining dollars can be divided by the total kWh to give you a cost per kWh. This cost will automatically include all surcharges, credits, taxes, etc. that are attributed to the energy consumption. Perform this calculation for every month and average the twelve values and you will have a good historical annual cost per kWh for your savings calculations.

The following computer spreadsheet shows you one format for an approach that requires only data entry from the electric bills and rates.

CUSTOMER NAME

BILLING DATA

MONTH/YR.	DEMAND RATE ($)	BILLED DEMAND (KW)	DEMAND COST ($)	TOTAL USAGE (KWH)	TOTAL COST ($)	KWH COST ($) / KWH
JAN. '93	$0.00	0.0	$0	0	$0	$0.0000
FEB. '93	$0.00	0.0	$0	0	$0	$0.0000
MAR. '93	$0.00	0.0	$0	0	$0	$0.0000
APR. '93	$0.00	0.0	$0	0	$0	$0.0000
MAY. '93	$0.00	0.0	$0	0	$0	$0.0000
JUN. '93	$0.00	0.0	$0	0	$0	$0.0000
JUL. '93	$0.00	0.0	$0	0	$0	$0.0000
AUG. '93	$0.00	0.0	$0	0	$0	$0.0000
SEPT. '93	$0.00	0.0	$0	0	$0	$0.0000
OCT. '93	$0.00	0.0	$0	0	$0	$0.0000
NOV. '93	$0.00	0.0	$0	0	$0	$0.0000
DEC. '93	$0.00	0.0	$0	0	$0	$0.0000

SUMMARY DATA

ANNUAL PEAK DEMAND	0.0	KW
DEMAND COST	$0	
KWH USAGE	0	KWH
TOTAL BILL	$0	
GROSS AVG. $/KWH	$0.0000	
INCREMENTAL $ PER KWH	$0.0000	

Examples and calculations:

ABC, INC.

BILLING DATA

MONTH/YR.	DEMAND RATE ($)	BILLED DEMAND (KW)	DEMAND COST ($)	TOTAL USAGE (KWH)	TOTAL COST ($)	*(a)* KWH COST ($) / KWH
JAN. '92	$9.00	1,251.2	$11,261	393,600	$36,900	$0.0651
FEB. '92	$9.00	1,209.6	$10,886	499,200	$43,345	$0.0650
MAR. '92	$9.00	1,152.0	$10,368	374,400	$30,563	$0.0539
APR. '92	$9.00	1,180.8	$10,627	435,200	$37,109	$0.0609
MAY. '92	$9.00	1,337.6	$12,038	460,800	$43,951	$0.0693
JUN. '92	$9.00	1,513.6	$13,622	534,400	$45,315	$0.0593
JUL. '92	$9.00	1,632.0	$14,688	412,800	$38,636	$0.0580
AUG. '92	$9.00	1,561.6	$14,054	534,400	$46,120	$0.0600
SEP. '92	$9.00	1,638.4	$14,746	566,400	$49,372	$0.0611
OCT. '92	$9.00	1,465.6	$13,190	502,400	$45,402	$0.0641
NOV. '92	$9.00	1,264.0	$11,376	473,600	$42,966	$0.0667
DEC. '92	$9.00	1,289.6	$11,606	460,800	$40,407	$0.0625

SUMMARY DATA

ANNUAL PEAK DEMAND	1,638.4	KW
DEMAND COST	$148,464	
KWH USAGE	5,648,000	KWH
TOTAL BILL	$500,086	
GROSS AVG. $/KWH	$0.0885	*(b)*
INCREMENTAL $ PER KWH	$0.0622	*(c)*

(a) (*Total Cost – Demand Cost*) ÷ *Total KWH*

(b) *Total Cost* ÷ *Total KWH*

(c) *Average Monthly "KWH Cost"* $\{\sum Col.\ (a) \div 12 = \$ / KWH\}$

Incremental Fuel Costs

Fuel units of consumption should be converted to Btu, or therms or mmBtu (million Btu). This will allow all sources of fuel to be averaged to develop a cost for the heat value purchased. The total cost per Btu will reflect the real costs relevant to the way the facility historically purchases fuel. If you also determine the various losses in combustion of each fuel and losses in the distribution of the thermal fluid system (the "end use efficiency"), you can develop an understanding of which fuel, including electricity, is the best choice for the specific application.

The following computer spreadsheets show you one format for an approach that requires only data entry from the fuel bills.

ABC, INC.

#6 FUEL OIL			*(b)*	0.15 mmBTU/GAL
MONTH/YR.	GALLONS DELIVERED	TOTAL COST	COST PER MMBTU ($/mmBTU)	
JAN. '91	0	$0.00	$0.00	
FEB. '91	0	$0.00	$0.00	
MAR. '91	0	$0.00	$0.00	
APR. '91	0	$0.00	$0.00	
MAY. '91	0	$0.00	$0.00	
JUN. '91	0	$0.00	$0.00	
JUL. '91	0	$0.00	$0.00	
AUG. '91	0	$0.00	$0.00	
SEP. '91	0	$0.00	$0.00	
OCT. '91	0	$0.00	$0.00	
NOV. '91	0	$0.00	$0.00	
DEC. '91	0	$0.00	$0.00	
			AVERAGE	
TOTALS	0	$0.00	*(c)* $0.00	

NATURAL GAS — 0.1 mmBTU/Therm

MONTH/YR.	USAGE (THERMS)	TOTAL COST	*(b)* COST PER MMBTU ($/mmBTU)
JAN. '91	0	$0.00	$0.00
FEB. '91	0	$0.00	$0.00
MAR. '91	0	$0.00	$0.00
APR. '91	0	$0.00	$0.00
MAY. '91	0	$0.00	$0.00
JUN. '91	0	$0.00	$0.00
JUL. '91	0	$0.00	$0.00
AUG. '91	0	$0.00	$0.00
SEP. '91	0	$0.00	$0.00
OCT. '91	0	$0.00	$0.00
NOV. '91	0	$0.00	$0.00
DEC. '91	0	$0.00	$0.00
			AVERAGE
TOTALS	0	$0.00	*(c)* $0.00

TOTAL FUEL; GAS & OIL

MONTH/YR.	OIL (GAL) DELIVERED	GAS (THERMS)	*(a)* TOTAL THERMS	TOTAL COST	*(b)* COST PER mmBTU ($/mmBTU)
JAN. '91	0	0.0	0	$0.00	$0.00
FEB. '91	0	0.0	0	$0.00	$0.00
MAR. '91	0	0.0	0	$0.00	$0.00
APR. '91	0	0.0	0	$0.00	$0.00
MAY. '91	0	0.0	0	$0.00	$0.00
JUN. '91	0	0.0	0	$0.00	$0.00
JUL. '91	0	0.0	0	$0.00	$0.00
AUG. '91	0	0.0	0	$0.00	$0.00
SEP. '91	0	0.0	0	$0.00	$0.00
OCT. '91	0	0.0	0	$0.00	$0.00
NOV. '91	0	0.0	0	$0.00	$0.00
DEC. '91	0	0.0	0	$0.00	$0.00
					AVERAGE
TOTALS	0.0	0.0	0	$0.00	*(c)* $0.00

(a) *(Gallons x mmBTU / Gal x (1 Therm / 0.1 mmBTU)) + (Natural Gas Therms)*

(b) *Total Cost ÷ (total Therms x (1 mmBTU / 10 Therms))*

(c) *Average Monthly Fuel Costs* $\{\sum Col.\ (b) \div 12 = \$ / mmBTU\}$

Chapter 2

Lighting

INTRODUCTION

I do not believe that I am a lighting expert. First as Plant Engineer, then as Corporate Energy Manager I have been exposed to many different options promoted by vendors. My decision as to whether to recommend implementation had to be based upon my own assessment of the validity. To be real honest, I had two main questions: 1) Would there really be a savings? 2) Will many people complain about the change?

Now, as an energy conservation consultant, I have made many lighting retrofit recommendations. I found that to perform with integrity I had to answer the same two main questions: 1) Would there really be a savings? 2) Will many people complain about the change? Much to my dismay, I found myself evaluating the claims of my competitors. This happens when you lose a bid. It also happens when a customer trusts you enough to let you know that your competitor has proposed to "save more energy for less money" but the customer doesn't believe it. He is faced with explaining to his boss why one is real and the other isn't.

In tough economic times this can be very difficult. Let's not mince words, Plant Managers and Accountants want the least cost option – period. If you want to spend more money you are the one they question, not the nice vendor offering a cheaper deal.

Every piece of technology has its place and can perform "as advertised". The vendor that thinks that his one product will be the best in all cases needs to be discovered and discarded. Use him where it's best. Use another where they are best. Or find a engineer that is not limited in his choices and force him to convince you that his choices are what's best for you.

New lighting products are capable of producing the best light possible for the energy consumed. But how much conservation is too much? When can a customer tell if the savings are too good to be true? The customer must rely upon the expertise of the vendor or engineer making the recommendations.

Unfortunately, too often, the person making the recommendations has his own interests at heart and is willing to sacrifice the long term interests of the customer.

The driving force for all energy conservation decisions is payback period... period. This necessarily brings up the subject of watts saved and burn time. Anyone who has tried to get an accurate determination of burn time knows that this is very difficult to get. The best solution is a negotiated settlement. The area of reduced watts is the real test of the vendor's integrity. Anyone can promise a watt reduction. The question is whether the customer can be convinced that he will be satisfied with the light levels after the job is finished. The vendor is armed with a confusing bag of terminology and subjective reasoning to boggle the most analytical mind. If the customer knew what the vendor was talking about then the customer probably would have done that retrofit long before the vendor showed on the scene.

Now mix in the confusing array of Utility Rebate Programs. Many of these programs are as good as they should be. They promote conservation and allow building owners to upgrade their facilities with the best technology. Unfortunately this means that they are considered money in the bank to vendors. Often the payment comes directly from the utility upon completion of the job. Certainly this accomplishes the favorable goal of removing the economic obstacle to conservation. However, it also removes the connection between customer satisfaction and payment.

It is in the vendor's best interest to maximize watts saved in order to assure that the annual savings are high enough to cover the net cost after the rebate. In other words, make the deal so sweet that it can't be refused.

DEFINITIONS

A wise (and remarkably unpretentious) college professor once told me that the only difference between a student and a Ph.D. is vocabulary. In order to make themselves indispensable, the typical lighting salesman will use a lot of lighting vocabulary. If you don't understand what he just told you, don't panic. My experience has been that *he* didn't understand what he just told you.

The really good lighting vendors truly appreciate an informed customer. It will make their job that much easier if they teach you the vocabulary, and they know it. So don't be afraid to let them know that you don't understand what they are trying to tell you. If they are not happy to take the time to teach you, tell them that there are others who would love to have the chance.

Here are some of the frequently used terms. You will see how difficult they can make them, and how easy they really are to understand.

Candela

One candela is the luminous intensity radiating from an area of 1/600,000 square meter of a blackbody radiator elevated to the temperature of solidifying platinum at a pressure of 101,325 Newtons per square meter.

Which means absolutely nothing to me. This is what I quote to people who think that the metric system is always the easiest system to work with.

Fortunately, one candela also happens to be approximately the light from one candle.

Lumen

A lumen is equal to the luminous flux through a unit-solid angle (steradian) from a uniform point source of one candela, it is also equal to the flux striking a unit surface all points of which are at a unit distance from a uniform point source of one candela.

Otherwise defined as a unit of luminous flux (light).

For the present, just understand that this is a small amount of light. A 60 watt incandescent is rated to generate 890 lumens, a single 40 watt fluorescent lamp is rated at 3,050 lumens.

Footcandle

A footcandle is a unit of measure defining the amount of light at a specific location. Its units of measure are: lumens per square foot.

As you can see, we have built up from candela to lumens to footcandle. Again, the definition may be intimidating as you work your way from candela to arrive at footcandles. Don't worry, only in the rarest circumstance will you run across candela in retrofit work.

The real world concept of footcandle is self-explanatory. Much like the lumen, it relates to a candle at a one foot distance. It is the amount of light at a specific location equal to the amount of light that would be there if the source was a candle held one foot away. Thus 50 footcandles is the amount of light that would be there if the source was 50 individual candles held one foot away.

Relative Light Output (RLO)

Relative Light Output is a new term that has emerged as more fixtures are being left in place and upgraded for efficiency. RLO is the heart of the retrofit conservation measure. It is the place for unscrupulous vendors to hide. And it is the place to expose them. Basically it is a percentage calculation comparing the amount of light from a given retrofit measure to the amount of light before any changes were made.

Unfortunately, now it gets vague. The question of how much light was there before changes were made brings up some new definitions that are relatively easy to understand, but are virtually impossible to quantify. Therefore the vendor can make the result almost anything he wants.

Fortunately, even though you can't pin it down, you can recognize when it's outrageous.

Initial

Whether this term is used with lumens or footcandles, it is referring to a new system, or one that has been cleaned and relamped with new lamps.

Depreciation

Light sources, i.e., lamps and fixtures, do not work like new all their lives. Lamps lose their brightness and fixtures lose their ability to reflect or transmit light. For lamps, the strict term is Lamp Lumen Depreciation, sometimes abbreviated as LD. For fixtures it's Lumen Dirt Depreciation (DD), which means exactly what it says. As the fixture gets dirtier, fewer lumens bounce off as more are absorbed.

These are all expressed as a percentage of the initial values.

Maintained

As it applies to lighting, this term refers to the level of light (percentage of initial) that will be experienced after several months have passed. The LD and DD can occur rapidly at first then level off for a long period of time. Thus the light level will remain relatively stable, or will be maintained unless the conditions change.

Their importance to our discussion lies in the fact that they are necessary percentages to calculate the condition of the existing system being evaluated for retrofit. This is where the reflector saves 50% of your electricity and gives you "the same light as before". The reflector vendors will tell you this, but they won't tell you that they mean the minute before you made the change and that they have assumed 80% LD and 80% DD.

So the real question is what is the light level that is required? You know what IES standards are. You know what certain other industry standards are. You might even know what light level they really need. Plant personnel will most often insist that more is better.

Coefficient of Utilization (CU)

The percentage of light from the lamps that escapes from the fixture that actually reaches the work surface. CU is a function of the fixture efficiency *and* the room characteristics.

Unfortunately, we don't like to see light bulbs. We enclose them and cover them and diffuse them. Every time that there is something between what you want to see and the light bulb, you lose some useful light. Please remember that this is necessary. Glare from bright sources can be so distracting that you cannot perform the task at hand.

Fortunately, for most retrofit options the CU is not changed and becomes a moot point. Only when you change something about the fixture or the room do you change the CU.

Color Temperature

Color temperature and the concept of cool and warm colors is a combination of the arts and sciences. Unfortunately that means that their usage is contradictory. The color temperature is the temperature, in degrees Kelvin, that a perfect (blackbody) radiator would have to be to appear the same color. A glowing ember is a good example. When it is just "red hot" the temperature is about 1800° K. A blue flame is closer to 6000° K. The contradiction comes in calling the lower temperature, red, a "warm" color and the higher temperature, blue, a "cool" color.

So why do you need to hear this? All you need to remember is that if you want to keep the same color rendition you need to start with the same color temperature. Two hundred degrees Kelvin either way will not make a significant difference.

Color Rendering Index

The term Color Rendering describes the effect a light source has on the appearance of colored objects. A light source has an absolute effect on how colors are perceived. If the light source produces light without a certain wavelength, blue for example, then blue will not be seen in any objects. The blue objects will appear to be black or gray.

The Color Rendering Index (CRI) is a rating from 1 to 100 that attempts to quantify the effect by comparing the way eight sample colors appear under the light source in question to the way they appear under an ideal light source (a blackbody radiator).

There are three things worth noting: 1) the CRI does not give you any idea whether some colors were better than others; 2) the comparison is only valid for comparing light sources operating at the same color temperature; and 3) I don't know who is looking at these colors and decides how good they are.

Unfortunately, most clients will not really understand the effects of light source and color. But, often they will insist that color rendition is important. Most of the time what they really mean is, "Don't change the colors that we are used to seeing."

If you don't want the colors to change try to use a new light source that is close to the same color temperature and the same, or higher, CRI.

Efficacy

Sometimes it is useful to know how well a light source converts electricity into light. This is calculated by dividing the lumen output by the watts consumed. The resulting quantity of lumens per watt is the called the efficacy of the light source.

Ballast Factor (BF)

The ballast factor is another measure of a system's performance. Unfortunately the ballast manufacturers chose a term that isn't descriptive so that there is no clue as to what it means. If they called it RLO you might not need them to explain. Basically, they compare the light output of the fluorescent lamp and ballast combination in question to the theoretical light output from a mythical perfect ballast. Thus a lamp and ballast combination with a ballast factor of 0.88 means that the light output will be equal to 88% of the rated lamp lumens in the lamp catalog.

Lamp Identification Nomenclature

If you want to confuse and intimidate the novice lighting customer you will rattle off the term "F40T12/CW/RS/WM" instead of saying "four foot fluorescent lamp". I admit that there is a lot more information in the first method, and it takes less time to say, so it is worth learning what it means. This is what my professor was talking about when he mentioned vocabulary making you smarter.

All lamps (the industry correct term for a light bulb is "lamp") are identified by a letter and number system. The letter defines the shape of the lamp. Sometimes the letter is descriptive, sometimes it is arbitrary. The number defines the physical size of the largest diameter of the lamp. The "T" in the example above stands for "Tubular", the "12" stands for the number of 1/8" in the diameter. I have no idea why it is or was important to know the diameter in eighths of an inch, but that's the way it is. So the lamp is a tubular lamp, 1-1/2" in diameter. The rest of the nomenclature is specific to identifying it as a reduced wattage, cool white, fluorescent lamp.

LIGHTING RETROFIT MEASURES

Lighting retrofit projects are relatively easy and can produce significant savings. With a little bit of information you can easily make a project go smoothly and profitably. When certain unscrupulous vendors sense that you have money to spend they will try to convince you that their product is the only thing that will work. Remember, when the only tool you have to sell is a hammer, everything looks like a nail.

What follows is some help in understanding how things really work.

Compact Fluorescent

This is an excellent conservation measure. The efficacy of fluorescent lamps is so much greater than incandescent that the savings potential can be as high as 87%! So what are the problems? The first, and often underlying problem, is size. The standard 60W A19 incandescent lamp has a maximum overall length (MOL) of 4-7/16" and a major diameter of 2-3/8". The lumen for lumen compact fluorescent is a 13w double or quad tube. The quad tube (or double twin), with a screw-in adapter has a MOL, depending on the manufacturer, closer to 6" with the major diameter around 2-3/4". The problem with the diameter is where it occurs. In an A19 lamp it is at the opposite end from the screw base. In the compact fluorescent and adapter it is right at the base. Most fixtures accommodate the A19 by virtually recessing the socket into the fixture body behind a reflector. In order to get a screw-in adapter to fit you usually need a socket extender close to 1-1/4" long.

The reason that I called this an underlying problem is that most of the time in a commercial/industrial setting the incandescent lamp is 75 or 100 watts. The lumen for lumen compact fluorescent for the 75w lamp is a 20w fluorescent with an MOL of 8-3/16". I don't know of a compact fluorescent that will replace the 1750 lumens of a 100 watt A19.

The whole reason for going over this is to enlighten you about the real and often encountered obstacles to utilizing this legitimate technological improvement. What you will often see is that the vendor will recommend 13w quads with adapters for all your hihats and porcelain sockets regardless of the existing wattage because that is pretty much the only one that will fit. This is fine if you are used to 60w incandescent lamps. Where it becomes a gray area is with decorative lighting, i.e., hihats and wall washers. In those cases there will be no loss of "productivity" or inconvenience if there is a reduction in the light level. Realistically, in many cases the only person who will get upset is the architect who is spending your money to make his "statement".

If you want to bite the bullet and replace the entire fixture there is another set of circumstances that are ultimately beneficial. The fixture manufacturers have gotten their acts together and designed new fixtures with the proper ventilation and optical reflectors that result in optimal output. Quite often these fixtures have a considerably higher CU than the existing incandescent fixtures and therefore require less lumens to produce equal RLO. In these situations you can reduce watts even further by reducing the lumen input.

That is what I mean about an informed decision to conserve. Only you can decide if the two questions (Will there really be a savings? Will many people complain about the change?) have been answered correctly.

As a final note on the subject, there have been recent studies on the effect of heat build-up on the light output from a compact fluorescent. Compact fluorescents have the same problems with lamp wall temperature (LWT) as all other fluorescents. Studies have shown that the LWT rises above design temperature when a compact fluorescent is enclosed in a hihat fixture or an integral reflector/lens. This has the overall effect of reducing the light output by as much as 20%. This, of course, makes the 13w compact fluorescent less than equal to a 60w incandescent lamp.

Fluorescent Current Limiters

These devices are connected to the ballast of a fluorescent fixture to reduce the lamp current. This is similar to what an electronic ballast does only it doesn't change the frequency. Therefore, whatever the reduction in power consumption there is a corresponding reduction in light output. However, with new lamps and a cleaned fixture, the RLO immediately after retrofit will very likely be acceptable. Is this a trick? Well yes and no. The lower lamp current will reduce the rate of lumen depreciation, therefore the light level will remain high longer. Many devices also clean up the sine wave of the lamp current, thus reducing the crest factor which has a large, positive impact on the lamp life and LD. It is a trick only when you are not told of the light loss or the reasons why the RLO should be acceptable.

One problem to mention: stroboscopic effect. My experience has been that reduced wattage lamps are very susceptible to exhibiting the sixty cycle flickering called stroboscopic effect. Very definitely eight foot slimline reduced wattage lamps are good candidates and *any* lamp that is exposed to cool air that will lower the LWT. This is easily seen in surface mounted strip fixtures that are located next to air conditioning diffusers.

Another problem: old ballasts. My experience has been that old ballasts (12 years or more) are susceptible to failure in the first month of use. I don't know why, though I have some guesses that it has to do with the ballast capacitor and resonance.

One last problem: low power factor. Measurements that I made on one brand indicated that they would contribute up to 50 VARs (lagging) per unit. This doesn't sound like much but we installed close to 14,000 of them in one plant so it made a big dent in their electric bill where they were being penalized $10/month for every KVAR. This problem has been addressed by the industry and has been corrected by many of the manufacturers.

Fluorescent Fixture Reflectors

The way reflectors are generally sold they are the embodiment of the phrase, "it's all done with mirrors". I mean that literally and in the sense that it's mostly hype. A new, baked white enamel fixture will have a reflectance factor between 82% and 85%. The best mirror reflector will have a reflectance factor of 96%. With only 14% improvement in reflectance, how can you get 50% more light out of the fixture. Admittedly, that 14% improvement represents a 17% increase over the original. And, it is true that the shape of the reflector is designed to get the light out of the fixture in one bounce. But both those conditions combine to produce an increase in the CU of the fixture of approximately 20%.

Don't get me wrong. I think that 20% is a significant achievement. But it is continuously oversold because it is also relatively costly and can be difficult to install.

Reflectors can be a useful tool in your kit when used properly. However; claims made by reflector salesmen ask you to believe that 50% of the light output from a fluorescent lamp is absorbed by the fixture. This is not so.

This kind of wild claim is easy to debunk when you know the simple dynamics of light reflections and the formulas for calculating fixture output. We have already discussed LD and DD. Common assumptions are that at the end of a fluorescent lamp's useful life the LD is 70%. The amount of DD is dependent upon the environment. In an office you might expect a range of 85% to 95%.

The vendor selling "the same light as before" is counting on large percentages of LD and DD. Percentages closer to 70% LD and 70% DD. Well, so what? The retrofit is complete and lo and behold the light level in the space is even a little better! The man is a hero. Better light for half the cost! But remember, you didn't change the location that caused the dirt and you didn't change the policy of relamping on burn out, so the same factors exist that made the 70% LD and 70% DD assumptions correct. Therefore; in about a year, the LD and DD will return and you will be back to 50% light output. But there's one big problem. You started with 50% light output! Fifty percent of 50% is only 25%. The only way out for a plant engineer is to maintain your system at its peak performance until you can get your resume back out on the street.

One more trick to expose. That first 100 hours (three weeks at one shift) that the new lamps were in place, and the evaluation was made, is only the lamp burn in period. New lamps experience a 10% lumen depreciation in the first 100 hours of use, then they level off. This is so well known that the lumen rating of a lamp that is published in the catalog and used in all footcandle calculations is really the rated output *after* 100 hours. The new lamps gave the vendor a nice 10% edge against complaints of low light level.

So how do you expose this? Ask him to work out the following equation for RLO:

% of existing initial lumens
x % increase in CU (or % increase in reflectance)
RLO

The typical numbers should be like this:

0.5 (1/2 the lamps)
x 1.2 (or % increase in reflectance)
0.6

The typical "real world" situation may have numbers that calculate out a little differently:

Before retrofit:

```
   1.0  (All the existing lamps)
 x 0.7  (LD)
 x 0.7  (DD)
 x 1.0  (existing CU)
 -----
  0.49
```

49% of the initial footcandles.

After retrofit:

```
   0.5  (1/2 the lamps)
 x 1.1  (LD of new lamp)
 x 1.0  (DD of clean fixture)
 x 1.2  (CU increase due to reflector)
 -----
  0.66
```

66% of the initial footcandles.

Which is a 35% increase over the existing conditions just prior to the retrofit. That sounds great! Half the watts with 35% more light. But look ahead ten or twelve months:

One year later:

```
   0.5
 x 0.7  (LD)
 x 0.7  (DD)
 x 1.2  (CU increase)
 -----
  0.29
```

29% of the initial footcandles.

Which is a 41% *decrease* in the amount of light present just prior to the retrofit.

Let me stress that I have seen *many* applications that can suffer a 41% light loss without any loss in productivity or safety. In hallways, for instance, this enables the retrofit to proceed without complaints. As time passes the light loss is gradual and unneeded, therefore unnoticed. But if the vendor is claiming "the same light as before" because you feel that you *need* the same light as before, be aware of how he will prove it.

Another "proof" will be the light meter test. The vendor will measure the light under a fixture before and after a reflector retrofit and, once again, it's at least the same if not better! Above I showed you how that can happen if there is poor maintenance and low LD and DD. But this will be the case, or very close, on even a well maintained system.

The reason is simple. Just look at any of those wonderful line drawings that reflector manufacturers include in their literature. There they show the rays of light radiating from the tube, hitting the reflector and bouncing down in one bounce. This is *exactly* what is going on and why the CU is increased. The reflector is designed to get the light out of the fixture in one bounce and it does it by directing the light downward. Therefore, although the total lumens are less, they are concentrated into a smaller square foot area so the lumens per square foot (footcandles) remains very much the same as before – *under the fixture.*

The best light meter test should include before and after readings taken carefully *under* the fixture and *in between* fixtures.

Again, I have recommended reflectors where this is a desirable effect, i.e., hallways, etc. But do it on purpose, not because you were tricked.

Lest I be accused of indiscriminate reflector bashing, let me identify a retrofit option that is not a hallway and is perfect for reflectors. A fixture that is a three lamp fixture with an eighteen cell deep parabolic lens is an extremely difficult fixture to retrofit any other way. If you just remove a lamp the row of cells below it will be very noticeably dark. However, a reflector and lamp socket relocation specifically designed for the fixture, can prove to be effective. The light is redistributed to all three cell rows from two lamp locations.

The RLO numbers work out like this:

0.667	(3 lamps to 2 lamps)
x 1.2	(CU increase)
0.8	

80% of the initial light output remains.

It is clear that if you want to factor even just a small amount of depreciation into the existing systems light level at the time of retrofit you will have equal or better light immediately after the retrofit.

Even better is the case where there are reduced wattage (34w) fluorescent lamps and reduced wattage ballasts in the existing system. This system starts with a ballast factor of 0.88. If the retrofit is to a two lamp T8 system (BF of 0.95), then the RLO numbers are these:

```
  0.667  (3 lamps to 2 lamps)
x 1.14   (lumen increase due to T8 and higher BF)
x 1.2    (CU increase)
-------
  0.91
```

91% RLO without hedging over depreciation.

Electronic Ballasts

I happen to be a true believer in electronic ballasts. The theory behind the use of high frequency to maintain the phosphorescent excitation at a higher level than at 60 hertz makes sense to me. I believe that it was a wise move on the part of the ballast manufacturers to choose the design goal of maintaining the existing light output and reduce the watts.

The industry has certainly gone through some turbulent times. Starting with the original design flaws that caused a high failure rate, overcoming the bad press from that, then dealing with demands for product that have been unbelievably high. Long lead times are a serious problem and frustration.

Dealing with these issues I have met and worked with some very nice, very dedicated people who manufacture and distribute this product.

Then there are the others.

The notion that one electronic ballast can operate almost any type of lamp (T8 or T12) is, as most tricks are, only partially correct. Fluorescent lamps are designed to operate at a specific lamp current.

The original purpose of a ballast was to limit this current to the design level and avoid the tendency for it to "run away", over pressure the tube, and cause damage.

Most T12 lamps operate at 425ma (slimline) or 430ma (4' or shorter rapid start). The 4' T8 lamps all operate at 265ma. In order to accommodate both, the manufacturers who try, designed the ballast to operate at a lamp current somewhere in between. As a result, the T8 lamps are over-driven and the T12 lamps are under-driven. The operating watts are a reflection of the lamp current. A two lamp T8 system driven by one of these electronic ballasts will use 76 watts. Only four watts less than a wattmiser lamp on a standard ballast. I have never heard this information "volunteered" by a salesman. They have told me when I insisted. To be fair, there is a considerable increase in light output over a wattmiser system. But over driving an expensive lamp to get more light is not sound engineering practice.

HIGH INTENSITY DISCHARGE LIGHTING

Although fluorescent lighting is technically high intensity discharge (HID) lighting, the term refers to Mercury Vapor (MV), Metal Halide (MH), High Pressure Sodium (HPS), and Low Pressure Sodium (LPS) light. Of those four types, LPS is the most efficient. Unfortunately, it has the color rendition that is the least acceptable. Although HPS light has a color rendition that may not be whole heartedly liked, it is widely accepted. Especially for outdoor use.

While MV and MH have the color rendering characteristics similar to fluorescent lamps, MV fixtures are not very efficient and MH fixtures are equal or slightly less efficient than fluorescent fixtures. A further drawback for MH fixtures is that the lamp life in wattages below 400 Watts is half that of HPS, MV and fluorescent lamps.

The problem is that fluorescents are pretty much limited to indoor use (poor low temperature performance) and fixture light output under 40,000 lumens. MH fixtures can have a light output up to 110,000 lumens. Plus the MH lamp is a virtual point light source that allows reflectors and lenses to direct the light.

Without a doubt, wherever there is a MV light it should be replaced. You should always try to install HPS lighting. Outdoor lighting should be easy to get approved with HPS. This is the highest efficiency and the longest lamp life. The failure mode is also self-correcting.

The failure mode of an MV lamp is to gradually become dimmer and dimmer until the light output is non-existent. (I have seen this in a dead storage warehouse. I thought all the lights were turned off until I saw one giving off a dull glow.) What this means is that unless you group lamp, you really don't relamp until it is painfully obvious. Especially an outdoor fixture that is in the back lot and 35 feet in the air.

The failure mode of a HPS fixture is to blink off and on until the neighbor's complain. What this means is that there is a real potential for reducing the lumens as well as the watts if you change from

MV to HPS. Why? Because the existing condition for years has probably been lamps with an average lumen depreciation of 70% or more! If this overall light output has been acceptable, then you can replace the MV fixtures with HPS fixtures and *not* size them lumen for lumen. You can at least downsize to 90% lumens or even 80% lumens without causing complaints about the light levels. (Complaints about the color are another subject.)

I, personally, do not recommend low pressure sodium lights for anything but the most basic security applications. It is my belief that it takes more than just lumens to allow effective "seeing". The monochromatic light grossly distorts all colors such that *contrast* and *perception* are greatly impaired. While objects become visible, they are difficult to identify without concentrating on them. This may be fine when the goal is to illuminate objects for safety and security. But if the tasks require identification and recognition, even if it's not too critical, the task becomes very difficult.

Occupancy Sensors

I have recommended and installed hundreds occupancy sensors. But evaluating the savings is a guess at best. Why? Because the savings are based upon correcting human errors. There is no easy way to accurately determine the level of human error for each office. All you can do is apply an assumed burn time percentage reduction.

This is dangerous at best. Your safest approach is to get the facility's personnel to make the guess for you.

When calculating the savings, don't double dip. Base the kWh reduction on reduced burn time for the wattage of the fixture *after* it is retrofit to the lowest watts. You may be surprised how poor the payback is under those conditions.

CALCULATIONS

All calculations are derived from the same goals. You must always remember: the first goal is to define the existing conditions. The second goal is to define the proposed conditions. The third, and final goal is to subtract the second from the first.

Defining the existing conditions is probably the most critical part. The calculations for a lighting retrofit are the easiest to handle in this business. The wattage before and after can be selected from a table of values for various fixture types. Since many utilities now have some kind of lighting rebate program, they have developed these tables to standardize the savings calculations. If you can't get such a table from your utility, then I suggest that you use the ANSI values from a ballast catalogue. It is important to use the fixture wattages since that will include all parasitic ballast consumption. The only other parameter required is burn time.

The following information must be gathered or guessed in order to arrive at an answer to the questions of economics:

Define the existing conditions:

1. Fixture watts before retrofit.
2. Quantity of fixtures.
3. Hours of operation (burn time).

Define the proposed conditions:

4. Fixture watts after retrofit.
5. Hours of operation (burn time).

Define the economics:

6. Cost to retrofit.
7. Incremental cost of electricity per kWh.
8. Incremental cost of electricity per kW, demand.

These are all reasonably self-explanatory. The only one to comment on is fixture quantity. The accuracy of the count should be inversely proportional to the quantity of the fixtures. It will make very little difference in your decision if the count you use is 3,500 instead of the actual 3,347, even though the error is greater than 150 fixtures. But if you are using 35 fixtures when the actual count is 25, you are in for a surprise.

The real goal is to determine the magnitude of the range of cost and savings. When you have an idea of how much money you are going to ask the accountants for, you know just how much time you need to spend answering their questions about your initial analysis.

The following computer spreadsheet shows you one format for an approach that requires only data entry from the walk through survey.

LIGHTING RETROFIT

ECM I.D.

EXISTING:	{Description}	**PROPOSED:**	{Description}
QTY:	0	**QTY:**	0
FIXTURE WATTS:	0	**FIXTURE WATTS:**	0
TOTAL BURN TIME:	0	**TOTAL BURN TIME:**	0
ENERGY COSTS:		**ENERGY CALCULATIONS:**	
INCREMENTAL kW:	$0:00	**kW SAVED:**	0.00
INCREMENTAL kWh:	$0.0000	**kWh SAVED:**	0
ENERGY COST SAVINGS:		**ESTIMATED INSTALLED COSTS:**	
kWh SAVINGS:	$0	**MATERIAL:**	$0.00
ANNUAL kW SAVINGS:	$0	**LABOR:**	$0.00
TOTAL SAVINGS:	$0	**CONTINGENCY:**	15%
		TOTAL/FIXTURE:	$0.00
PAYBACK PERIOD:	0.0	**TOTAL ECM:**	$0

{Example and Equations}

LIGHTING RETROFIT

OFFICE FLUORESCENTS

EXISTING:	4'-3LT12ES/ES*	PROPOSED:	4'-3LT8-EB/LP**
QTY:	278	QTY:	278
FIXTURE WATTS:	124	FIXTURE WATTS:	78
TOTAL BURN TIME:	4431	TOTAL BURN TIME:	4431
ENERGY COSTS:		**ENERGY CALCULATIONS:**	
INCREMENTAL kW:	$9.00	kW SAVED:	*(a)* 12.79
INCREMENTAL kWh:	$0.0565	kWh SAVED:	*(b)* 56,672
ENERGY COST SAVINGS:		**ESTIMATED INSTALLED COSTS:**	
kWh SAVINGS:	*(c)* $3,202	MATERIAL:	$32.50
ANNUAL kW SAVINGS:	*(d)* $898	LABOR:	$20.00
TOTAL SAVINGS:	*(e)* $4100	CONTINGENCY:	15%
		TOTAL/FIXTURE:	*(f)* $60.38
PAYBACK PERIOD:	4.1	TOTAL ECM:	*(g)* $16786

* {4'; 3-Lamp; T12; reduced wattage lamps/ reduced wattage ballast}

** {4'; 3-Lamp; T8; electronic ballast/low power design}

(a) { [(Exist. Qty.) x (Exist. Watts)] – [(Prop. Qty) x (Prop. Watts)]} ÷ 1,000

(b) { [(Exist. Qty.) x (Exist. Watts) x (Exist. Hrs.)] – [(Prop. Qty.) x (Prop. Watts) x (Prop. Hrs.)] } ÷ 1,000

(c) kWh Saved x Incremental $ / kWh

(d) kW Saved x Incremental $ / kW x 12 Months

(e) kWh $ Saved + kW $ Saved

(f) (Mat'l $ + Lab. $) x (1 + Contingency)

(g) Total $ / Fixt. x Prop. Qty.

The following spreadsheet offers an approach to Occupancy Sensors that can be used for the entire facility before or after the lighting retrofit.

OCCUPANCY SENSOR

ECM I.D.

PROPOSED:		EXISTING	
SENSOR QTY:	0	BURN TIME:	0%
		% BURN TIME REDUCTION:	0%
DESCRIPTION:	{Description}		
FIXTURE QTY:	0		
FIXTURE WATTS:	0	TOTAL KW:	0.00
DESCRIPTION:	{Description}		
FIXTURE QTY:	0		
FIXTURE WATTS:	0	TOTAL KW:	0.00
DESCRIPTION:	{Description}		
FIXTURE QTY:	0		
FIXTURE WATTS:	0	TOTAL KW:	0.00
DESCRIPTION:	{Description}		
FIXTURE QTY:	0		
FIXTURE WATTS:	0	TOTAL KW:	0.00
		TOTAL SENSOR KW:	0.00

ENERGY CALCULATIONS:		ESTIMATED INSTALLED COSTS:	
INCREMENTAL kWh:	$0.0000	MATERIAL:	$0.00
		LABOR:	$0.00
KWH REDUCTION:	0	MISC.:	$0.00
kWh SAVINGS:	$0	CONTINGENCY:	15%
		TOTAL SENSOR:	$0.00
PAYBACK PERIOD:	ERR	TOTAL ECM:	$0

{Example and Equations}

OCCUPANCY SENSOR

ADMIN. OFFICES

PROPOSED:			
SENSOR QTY:	30	EXISTING BURN TIME:	2860
		% BURN TIME REDUCTION:	25%
DESCRIPTION:	4'-4LT8-EB		
FIXTURE QTY:	50		
FIXTURE WATTS:	106	TOTAL KW:	***(a)*** 5.30
DESCRIPTION:	4'-2LT8-EB		
FIXTURE QTY:	25		
FIXTURE WATTS:	62	TOTAL KW:	***(a)*** 1.55
DESCRIPTION:	{Description}		
FIXTURE QTY:	0		
FIXTURE WATTS:	0	TOTAL KW:	***(a)*** 0.00
DESCRIPTION:	{Description}		
FIXTURE QTY:	0		
FIXTURE WATTS:	0	TOTAL KW:	***(a)*** 0.00
		TOTAL SENSOR KW:	6.85
ENERGY CALCULATIONS:		**ESTIMATED INSTALLED COSTS:**	
INCREMENTAL $/kWh:	$0.0650	MATERIAL:	$50.00
		LABOR:	$15.00
KWH REDUCTION:	***(b)*** 4,898	MISC.	$0.00
kWh SAVINGS:	***(c)*** $318	CONTINGENCY:	15%
		TOTAL SENSOR:	$74.75
PAYBACK PERIOD:	7.04	TOTAL ECM:	$2,242

(a) [(*Fixture Qty.*) x (*Fixture Watts*)] ÷ 1,000

(b) $\sum$*(a) * Burn Time * % Burn Time Reduction*

(c) *kWh Saved* x *Incremental $ / kWh*

ANECDOTES

Richard's Retrofit Rules

Lighting Rule #1: *"You can never save more energy than shutting it off."*

Lighting Rule #2: *"There is at least one person who won't like it."*

Lighting Rule #3: *"Always retrofit lighting at night."*

Lighting Rule #4: *"The occupancy sensor will turn the lights off when the company president is in the bathroom."*

I recently completed a lighting retrofit project for a major office building, forty stories of luxury office space. Being very careful, I chose to play it safe and replace the incandescent hihat fixtures with new compact fluorescent hihat fixtures. To be certain of success, I chose replacement fixtures with equal, or greater, lumen output. Fortunately, I insisted on installing a sample set of fixtures in the pickiest tenant's space.

It was a lawyer's office. Four floors of the forty, 10% of the rentable space, controlled by this one law firm. They had clout and I knew it, so I wanted them to sign off on the selection. Essentially, the building manager agreed that if they were with us, the other tenants would not complain.

The test area was one of the break rooms; six fixtures, each with two 13W compact fluorescent lamps, in a ten by twelve windowless room. The only other light came from the fully lit soda machine door and the snack vending machine. I thought it looked great.

The firm's Facility Coordinator thought that it was acceptable. I still waited a week before I ordered 1,960 fixtures to complete the project. I should have waited two weeks.

"Marilyn thinks it's too harsh and too bright." I quickly put the 700 fixtures for their floors on hold. I never found out what Marilyn had on everybody, but they were all worried about her feelings. She didn't have a corner office, but maybe there was some significance to the fact that her office was *between* two corner offices. After two more tries I got the best response from Marilyn that any retrofit engineer can get: "I couldn't find the fixtures that were retrofit."

I now ordered 700 fixtures with two 9W compact fluorescent lamps each. Installing fixtures at a rate of 40 fixtures a night, we had 240 installed before the Facility Coordinator asked us to stop because the Art Committee (I'm not making this up) was "afraid that the paintings on the walls would look different."

We installed several wall washers so that they could see the effect. Fortunately, they couldn't agree whether the change was better (more light on the paintings) or worse (slightly different color rendition). We were allowed to proceed.

Approximately 200 fixtures later, someone noticed that the brass trim was not as wide as the old fixture. Once again installation stopped. My first response to the building manager was to advise him to take a hard line. Since they had over six weeks to notice the difference and had given the OK to proceed, they should bear the cost to remove and return the narrow trim and the increased cost of the wide trim. My second response was to advise him to stall for a week. They would have to go back to work sometime and stop looking at the ceiling.

Fortunately, faced with a $12,000 charge and a workforce wasting far too much time looking at the ceiling, Marilyn said that the trim was OK. The installation finally went full speed ahead.

I learned the lesson about installing lighting retrofit measures at night when I was a corporate energy manager. I was implementing a fluorescent lighting retrofit project for the home office facility. To avoid interfering with personnel I arranged for the contractor to work nights after everyone went home.

One morning, a few days into the project, I was confronted in my office by a lynch mob. "We cannot work under these conditions!" "We are all getting terrible headaches!"

Suddenly I felt very, very lucky. I was lucky because I hadn't retrofit their area yet! Not only did it shut them down, it shut down anyone who had a legitimate complaint! No one wanted to take the chance of looking *that* foolish again. I was lucky because, while it was obvious in this case that their complaints were completely unfounded, if I *had* retrofit their area I wouldn't have an argument to counter with. It would have been a power struggle just to complete the project.

Another story involves the fact that I strongly believe that you should always try to live with retrofit measures that you are recommending for other people. So I installed an occupancy sensor in my office/study at home. With only about 150 watts of various incandescent lights it certainly isn't cost effective, but it provided me with invaluable experience of what it is like to live with one.

I really had no trouble putting up with the lights going off while I was working at the computer, or reading technical manuals, or talking on the phone. I increased the time delay and reduced the number of these annoyances (and reduced the savings potential). I got used to waving my arms or rocking in my chair to put the lights back on (and reduced the life of the lamps). I even put up with the faint hum on the telephone. I found that I could eliminate the hum by shutting the lamps off by their switches during the day (and again reduced the savings potential).

But there was one thing that I could not tolerate.

One night, in bed, snuggling with my wife, I noticed that she was tense and unresponsive. When I asked what was wrong, she answered: "I hate that damn switch in your study!" (I'm not making this up.) "When I go in there during the day it turns the lights on when I don't want them on. When I'm talking on the phone at night it turns the lights off when I don't want them off. I have to flap my arms like some stupid bird to get the lights to work! That switch controls me! I want to control *it!*"

It took me less than ten minutes to put the standard light switch back in the study. It may be "my" office, but this house is *her* domain. And I will not tolerate anything that interferes with my "snuggling" with my wife.

Don't forget: the customer probably got to be president of the company because he is a control junkie. If he doesn't have control of everything in his environment he is extremely uncomfortable.

Chapter 3

Pumps

INTRODUCTION

Most inexperienced energy engineers discover the Affinity Laws and go crazy. The potential for energy savings is almost unbelievable. In the simplest situations, the savings are absolutely true. The true test is to be able to recognize when things are not simple and the evaluation needs to be modified.

Rest assured, the modifications all follow the Affinity Laws, the calculations are simple, and the savings are usually considerable. But the savings potential will be less, and in some cases too small to consider when certain easily recognized situations exist.

At the time of this writing, the hottest item in "pumping energy conservation" is variable speed drives (VSD). As always, a good engineer will recognize those situations where it will be more cost effective to downsize the pump or change the operating strategy. But these situations will be discovered while you are evaluating the potential for a VSD application. Pumps that are never required to run at full speed are a dead give away for downsizing. Remember, you can always put a VSD on a downsized pump to optimize the retrofit costs *and* the savings.

DEFINITIONS

Affinity Law

I had to look up the word "affinity" in the dictionary to get an idea why it was chosen for the formulas defining the relationships between speed ratios, head ratios and horsepower ratios for centrifugal pump and fan applications. I found out that the word "affinity" means "relationship". It is my feeling that a deliberate effort is made every now and then to name useful and clever formulas with words that tend to hide the connection to their use in the real world. Possibly it is just a result of engineers taking a cue from doctors and lawyers.

Anyway, the ratio equations are straight forward and simple to manipulate to determine any useful information with a few known facts:

$$\frac{N_1}{N_2} = \frac{F_1}{F_2}; \quad \left(\frac{N_1}{N_2}\right)^2 = \frac{H_1}{H_2}; \quad \left(\frac{N_1}{N_2}\right)^3 = \frac{HP_1}{HP_2}$$

N = rpm F = Flow H = head HP = horsepower

Head

With respect to fluid flow, this term is synonymous with the word "Pressure". By far it most often refers to pressure given in units of Feet of Water Column, other times it may be Pounds per Square Inch.

Static Head

The static head is the pressure at the pump caused totally by the highest elevation of the water above the pump (which is why the units often are in feet). It is the pressure developed or required even when there is zero flow in the system (static conditions). In a closed loop the static head at the outlet of the pump must be the same as the static head at the inlet to the pump. Therefore the static head that the pump must overcome to move water in a closed loop is zero.

However, even in a closed loop the minimum pressure required by the system header piping to provide the necessary flow through all connected devices should also be considered to be the static head of the system. If, for example, you are supplying cooling water to several parallel fan coils designed for operation with a 30 psi inlet pressure, the static (or zero flow) head that the pump must deliver into the piping system is the 30 psi (approximately 70 feet). Even though you could certainly let the header pressure go to zero when no flow is required, for all practical purposes it is better to maintain the required pressure at all times. Why? Because 1) there will be no complaints if the system is maintained such that it is capable of responding instantly to a call for flow, and 2) in most cases in the real world the probability of all connected devices being at zero flow is very small.

Friction Head

The friction head is the pressure losses due solely to the friction of the water moving against the pipe and fittings. (If you have started with units of feet, remember to use the same units in all your calculations.)

Pressure Drop (PD)

The pressure drop is the change in pressure between the inlet of something and the outlet of something. In the simplest system of a closed loop without any equipment, the pressure drop from the beginning to the end is the friction head. In the next simplest system of a fluid transfer arrangement, like from a sump to the top of a cooling tower, the pressure drop from the beginning to the end is the friction head plus the static head.

All pieces of equipment that require cooling or heating water will specify what the pressure drop through the equipment is expected to be when you operate it at its rated heat exchange level. This is the friction head that needs to be overcome in order to push water through the equipment at the flow rate for which it is designed.

If equipment is in parallel, the equipment with the highest PD is the pump controlling factor. If the equipment is in series, the sum of each PD is the pump controlling factor.

Total Dynamic Head (TDH)

The TDH is the highest pressure that the piping and equipment system will impose on the pump. This is simply a sum of the static head, the friction head and the controlling pressure drop of the various pieces of equipment.

WHAT TO LOOK FOR

Centrifugal pumps used to circulate cooling (or heating) water to HVAC or process applications offer major savings potential. Look for throttling valves. Not just at the outlet from the pump, but at the various end use equipment. Many process cooling water loops are supplied by a pump that runs constantly pressurizing a plantwide header with various machine taps ending in a valve (solenoid or manual) at each machine. This is schematically equivalent to having a throttling valve on the pump outlet that is actuated in discrete steps.

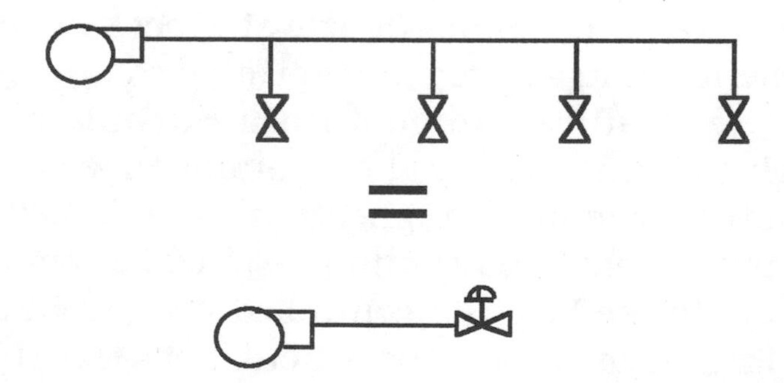

Each of those machine valves varies the flow requirement from the pump by an incremental amount. I usually recommend a variable speed drive

(VSD) controlled by a pressure sensor in the header. As a valve opens, the flow increases and the pressure drops. The pressure sensor will send a signal to speed up the pump to maintain the pressure setpoint. As a valve closes, the flow will decrease and the pressure will rise, slowing the pump down.

This measure should also be evaluated for HVAC chilled water loops with two-way, modulating valves at each coil. If the existing system uses three-way modulating bypass valves, the savings will be optimized if the valves are replaced with two-way valves or modified with the bypass isolated and capped.

Cooling towers transfer heat from the inside of a facility to the outside. When used with refrigeration equipment the heat is transferred from the condenser side of refrigeration machines to the outside air. When used with process equipment the heat is transferred from a process heat exchanger or cooling jacket to the outside air. Installing a VSD controlled by return water temperature will ensure that the gpm (mass flow rate) is matched to the Btu rejection requirements. (**Caution!** Check with the equipment manufacturer to determine the minimum flow rate required to maintain turbulent flow through the condenser and evaporator tubes.)

Another common application is a boiler feed water pump that runs continuously while the actual inlet flow to the boiler is controlled by a modulating (throttling) float valve. The calculations must take into account the fact that the pump must provide a minimum pressure higher than the boiler pressure. This means that there is a minimum horsepower requirement even if the flow is negligible.

Therein lies the difficulty. There are *many* applications where the minimum speed of the pump will be dictated by the minimum TDH. The trick is to recognize that there *is* a minimum TDH and then determine what it is.

All pumps are selected to operate at a certain maximum design flow rate, F_D and maximum design total dynamic head; H_D (D=Design). H_D is the total of static and friction head. (Fig. 1).

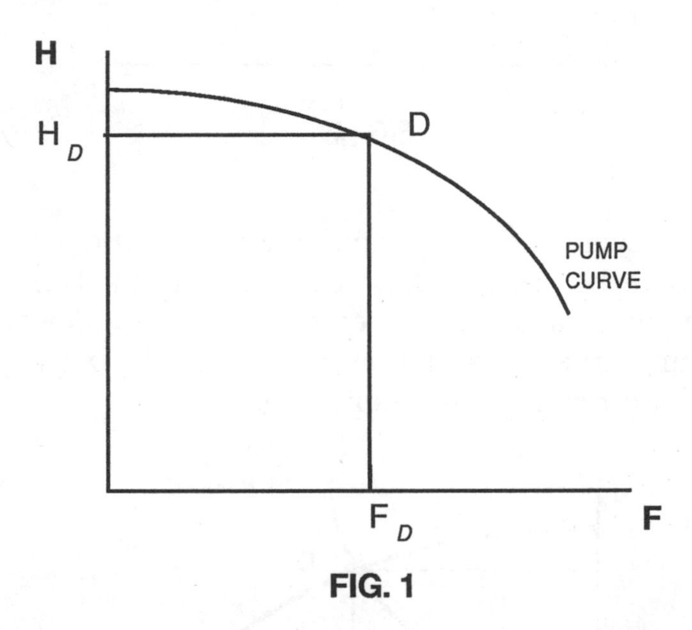

FIG. 1

For each system there may be a specific minimum static head requirement, $H_{Min.}$ If there is no minimum static head, such as a closed loop for hot water baseboard heat, then the system curve will start at zero head at zero flow. The friction head, however, varies as the square of the flow rate (Fig. 2).

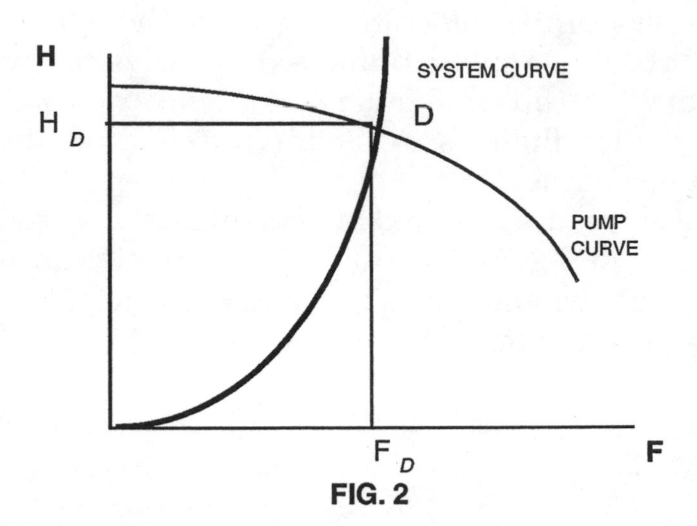

FIG. 2

However, most systems have a static head to overcome before *any* flow can be started. Since the static head will remain constant for all flows, the system curve does not start at zero H because of the static head requirement (Fig. 3).

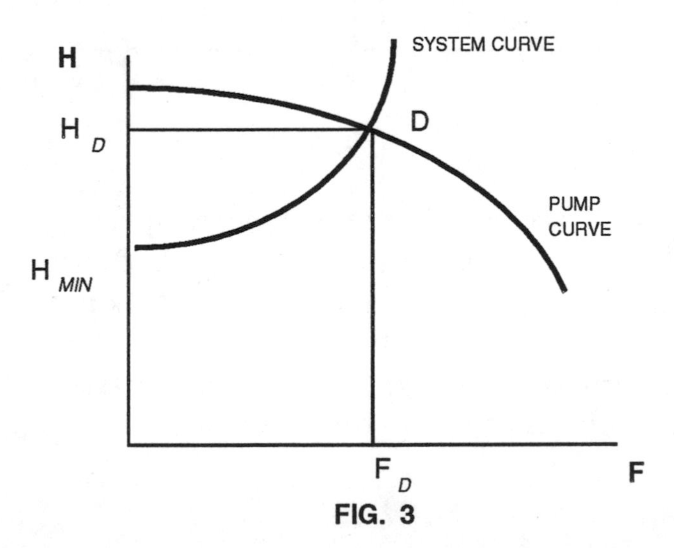

FIG. 3

To use a VSD means that you are developing the "family" of pump curves, i.e., a curve for each RPM. The minimum RPM curve (RPM 4) is the one closest to the minimum static head (Hmin) at zero flow (Fig. 4).

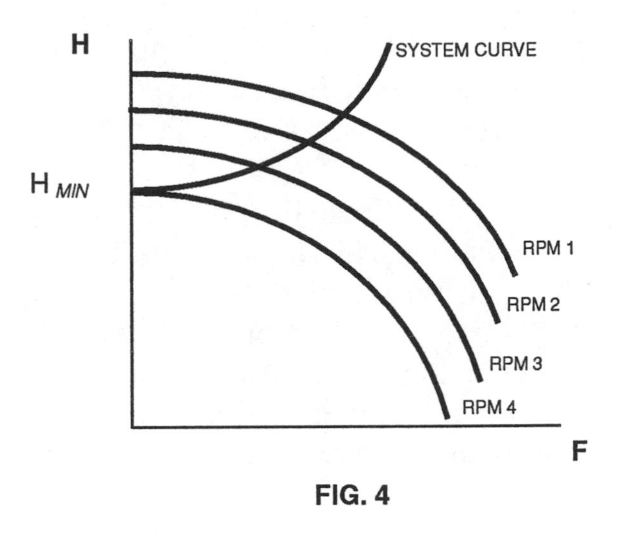

FIG. 4

That minimum RPM curve can be found by the Affinity Laws. The same laws identify the minimum horsepower requirement. Potential energy savings will be available between (H_{min}) and H_D.

$$\frac{N_{min}}{N_D} = \frac{F_{min}}{F_D}; \quad \left(\frac{N_{min}}{N_D}\right)^2 = \frac{H_{min}}{H_D}; \quad \left(\frac{N_{min}}{N_D}\right)^3 = \frac{HP_{min}}{HP_D};$$

$$\left(\frac{N_{min}}{N_D}\right)^2 = (\%SPEED)^2_{MIN} = \frac{H_{MIN}}{H_D}$$

$$(\%SPEED)_{MIN} = \sqrt{\frac{H_{MIN}}{H_D}}$$

Therefore:

$$HP_{MIN} = HP_D * \left(\sqrt{\frac{H_{MIN}}{H_D}}\right)^3$$

In the real world, the amount of pump horsepower savings is determined usually by the amount of friction head variation caused by the variation in flow rates. The greater the friction losses at high flow as compared to low flow the greater the horsepower savings available. The best situation is one that will also have a reduced pressure requirement. However; this is also a situation that should be evaluated for downsizing the pump first. The application of a VSD to a smaller pump results in an optimization of cost and savings.

THINGS TO LOOK OUT FOR ARE:

1. *Constant high flow requirements:* If there is no variation in flow and the pump is maxed out, there is no place to save energy. If, however, there is no variation in flow and the pump is grossly under loaded, first try to change the pump to the right size. If operating personnel can't part with their existing equipment (see Pump Retrofit Rule #2), then the installation of a VSD will save a lot of money.

2. *Non-variable fluid transfer systems:* In many process pump applications the goal is to move a quantity of fluid from one place to another as fast as possible. The size of the pump and motor are

determined by first cost economics. If they could afford to buy a 100 horsepower pump and reduce the transfer time by 50% they probably would. There is no potential for savings because the process requires the short burst of consumption and then the pump is shut off (see Pump Retrofit Rule #1).

This is the situation with municipal water supply systems. The district water distribution and booster pumps are generally controlled by a float switch in the storage tank. As the float drops to the low level setting, the pump is started and runs flat out until the float high level is satisfied and the pump is shut off. It may be argued that an analog signal from the float could be used to vary the speed of the pump to match the flow demanded from the system. But remember that the static head in these storage tank systems is usually quite high requiring a relatively high minimum speed. In addition, I have learned that the action of the large changes in water level has the desired effect of breaking the ice that forms on the surface of the water.

3. *Positive displacement pumps:* The use of a VSD on a positive displacement pump is not a problem. The only reason to be aware of such pumps is that they do not necessarily follow the affinity laws. The affinity laws apply to centrifugal loads. Centrifugal loads exhibit the characteristic of requiring variable torque with the variable flow. Positive displacement loads require constant torque over the total range of flows. The horsepower to speed relationship can be a straight line at worst or a square law relationship. Either way saves energy, but the

calculated savings will be less than with a centrifugal load that follows the cubed relationship.

Typical examples of positive displacement loads are piston type pumps for highly viscous fluids and mechanical aerators (paddle wheels) for aeration lagoons.

CALCULATIONS

Remember, you have two goals:

1. Correctly define the existing conditions in terms of total energy consumption.

2. Estimate the total energy consumption that will occur *after* the project is implemented.

The first goal is simply the horsepower of the pump motor times the hours it runs.

The second goal is simply the horsepower of the pump motor at reduced speeds (flows) times the hours that it runs at those reduced speeds.

If you have the system design documents with system and pump curves, or a table giving the gpm that the system requires at the TDH it was designed for, then your problem is *very* simple. The minimum and design pressure will be known so that you can calculate the minimum pump speed for the system, i.e., where the system curve intersects the zero pressure axis. The Affinity Laws will define the speed percentage for the flow portion of the system curve. Remember:

$$\frac{N_1}{N_2} = \frac{F_1}{F_2}; \quad \& \quad \sqrt{\frac{H_{MIN}}{H_D}} = \% \text{ Min Speed}$$

N = rpm $\quad F$ = flow $\quad H$ = head

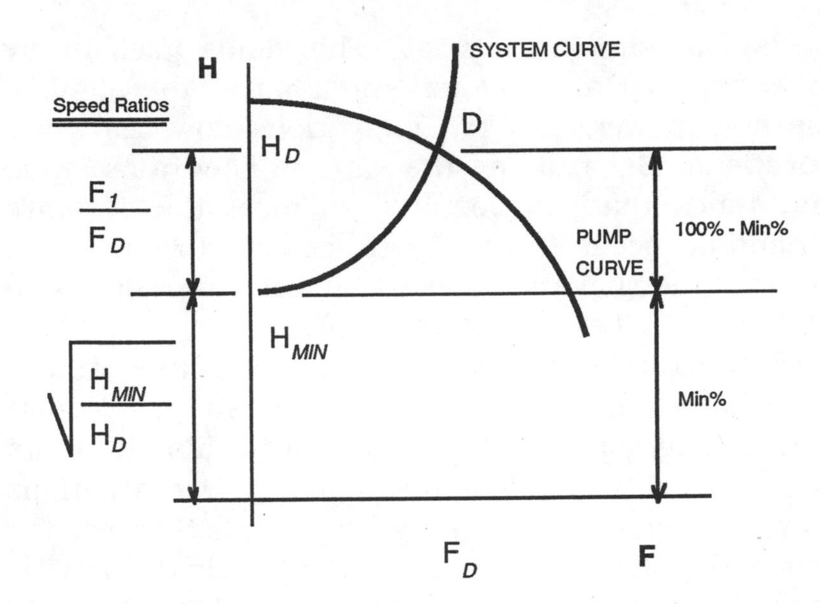

Therefore, the equation giving the speed percentage for any flow is:

$$\% Speed_1 = Min\% + \left[\frac{F_1}{F_D} * (100\% - Min\%) \right] =$$

$$\frac{N_1}{N_D} = \sqrt{\frac{H_{MIN}}{H_D}} + \left[\frac{F_1}{F_D} * \left(1 - \sqrt{\frac{H_{MIN}}{H_D}} \right) \right]$$

Finally, the horsepower required for any flow, F_1, is the Affinity Law:

$$\left(\frac{N_1}{N_D} \right)^3 * HP_D = HP_1 = \left\{ \sqrt{\frac{H_{MIN}}{H_D}} + \left[\frac{F_1}{F_D} * \left(1 - \sqrt{\frac{H_{MIN}}{H_D}} \right) \right] \right\}^3 * HP_D$$

$N = rpm \quad HP = horsepower$

Is that simple, or what? This lends itself nicely to setting up a computer spreadsheet to calculate the horsepower, and kW, for various flow rates and durations. Determining the various flow rates (gpm) and times that are required to meet the system's functional needs is what you get paid to do. If you didn't think that there would be any variations then you wouldn't be evaluating a VSD.

Unfortunately, for the majority of cases that I have had to deal with, the system design information is long gone. When you run into this situation: don't panic. The calculations for these evaluations may, at first, seem anything but simple. Indeed, the concepts behind the equations and their derivations can be intimidating. However, you don't have to derive the formulas every time you use them. And, while it is always better to understand the finer points of an internal combustion engine*, you don't need to be a mechanic to drive a car. Certainly driving the car without knowing the rules of the road can be very dangerous, and just driving around without knowing the route can land you someplace that you don't want to be. But going slow at first, in familiar territory, by yourself, can be very productive *and* give you valuable experience before you have to deal with a super highway in rush hour.

All you need are three equations.

To find the horsepower at *any* flow you simply have to fill in the unknowns in the following equation:

(1) $Horsepower = (GPM * TDH) \div (3960 * Pump\ Efficiency)$

* Even with degrees in electrical and mechanical engineering, I firmly believe that internal combustion engines operate on the principles of black magic. As evidence I offer the following observation: anyone who works on them must resort to chanting magical words, usually containing four letters and delivered at high volume.

To determine the TDH you can use the next two equations:

(2) *TDH = Static Head (ft.) + Friction Head (ft.)*

(3) $$\text{Friction Head (ft.)} = \left(0.001246 * GPM^2 * \text{Equivalent Lineal Feet of Pipe}\right) \div (\text{Pipe Diameter in inches})^5$$

(Note: the constant 0.001246 is the factor for galvanized or black iron (rough) pipe. For brass, copper, or plastic (smooth) pipe, the factor is 0.000623.)

Combined the horsepower equation looks like this:

$$\text{Horsepower} = \frac{\left(GPM * (0.001246 * GPM^2 * \text{Equivalent Lineal Feet of Pipe}) \div (\text{Pipe Diameter in inches})^5\right)}{(3690 * \text{Pump Efficiency})}$$

The object is to input all *known* parameters and then manipulate the *unknown* parameters in the horsepower equation (design gpm and TDH) to arrive at an equation that results in a calculated full load horsepower consistent with the nameplate on the motor and the selected motor load factor. From that point on you can vary the gpm in the equation to determine the horsepower at the different flows.

The following information must be gathered (or guessed) in order to arrive at an answer to the question of TDH:

1. Static Head.

This parameter should be the easiest to determine. If there are no gages, the facility personnel should know the pressure requirements and/or the physical elevation changes of the systems involved. If available, use the original construction plans and specifications. If all else fails, look through equipment specifications for similar equipment. You can at least base your assumptions on realistic values from someone's published data.

2. Equivalent lineal feet of pipe.

This is an undeniable wild guess. If the system is a plant wide cooling loop, use the perimeter of the area served. In any case, don't forget to figure on a round trip for a circulating system. A good estimation of the effect of valves and fittings is to take the estimated distance and multiply it by 2.5 to get the equivalent lineal feet. Get a feel for the complexity of the run. At a city water supply facility it was more appropriate to use a 1.5 multiplier because the pumps served several miles of relatively straight pipe.

This parameter is, however, a value that can, and should, be "tweaked" to establish the estimated system curve. Once all the other data is satisfied, making realistic changes in this value can bring the horsepower calculated at your peak flow gpm in the spreadsheet to match the installed motor.

3. Average pipe diameter.

Always start off with the header diameter at the pump outlet. This will be most accurate for short runs such as boiler feed water systems. If the resulting peak horsepower in the spreadsheet is unrealistically low, adjust this value downward to a realistic effective average diameter that will reflect the effects of longer branch systems with smaller diameter pipes in distant branches.

If you followed closely, you will note that you have solved the two steps. The energy before was the pump horsepower running at a constant load all the time. The energy after is the pump horsepower running at various gpm for various periods. Just run the horsepower equation for the various flow rates and you will know how much energy it will take to run the system with a VSD. Subtract the after energy from the before energy and you know how much energy you will save.

I told you it was simple.

The following information must be gathered or guessed in order to arrive at an answer to the questions of economics:

1. Active pump horsepower.

The key word is "active". The pump has to be running to save energy. This may sound obvious to some, but if you find a duplex pumping system that is alternated, both pumps need to be controlled but only one uses energy at a time.

2. Motor efficiency.

If the efficiency isn't stamped on the nameplate then make a logical assumption (guess).

3. Assumed load factor.

No motor is loaded to 100%. If a design engineer calculates the horsepower required to be 4.5 horsepower he won't select a 5 horsepower motor (if he's smart). Why? Because he knows that the chances of his design information being that correct are slim.

He should select a 7.5 horsepower motor. If you come across a motor that *is* fully loaded, you have also come across a situation where there is no potential for conservation.

If you assume the existing condition utilizes the full nameplate horsepower you will be making the all too common error of overstating the savings potential. If I don't have anything else to base my assumptions on, I use a load factor of 80%.

4. Pump run time.
5. Assumed normal "OFF" time.

These may sound like a redundant way to say the same thing. What I am addressing is the situation where the controls are automatic. Many times all the facility personnel will be able to tell you is that the pumps are "enabled" during certain time periods. You will have to develop some level of understanding about the control strategy so that you can arrive at an assumed percentage of the enabled ON time that the pumps will be automatically turned OFF.

6. Maximum gpm.

Most of the time this will be given. If you find that you have to guess, you can start with the formula:

$$GPM = 15 * D^2$$

D = Pipe Diameter; inches
15 = Flow velocity factor;
{15 = avg; 10 = conservative; 20 = borderline}

Again, if available, use the original construction plans and specifications.

7. Pump efficiency.

For gpm ranges less than 100 gpm use 70% to 75%, for 100 to 500 gpm use 75% to 80%; for large pumps: 85%.

8. Operating scenario; percent time at percent flow.

I usually use four steps for percent flow: 100%, 85%, 75%, & 65% (each for an equal 25% of the time) for circulating pumps that are not used in comfort heating or cooling. As stated in previous chapters, try to get the facility personnel to make these assumptions for you. They should have a better feel for what's right and wrong. (This may sound like it contradicts Retrofit Rule #3. It does. But unless you have better data than they do, let *them* make the guess.) Don't forget to check the minimum speed requirement for the system parameters. You don't want to look like you don't know what you're doing. Especially if you don't.

The following spreadsheet offers an approach that can be used for circulating pumps.

CIRCULATING PUMP VARIABLE SPEED CONTROLLERS
CIRCULATING PUMP IDENTIFICATION

BASE ASSUMPTIONS:

ACTIVE PUMP HORSEPOWER:	0
ASSUMED EFFICIENCY:	0%
ASSUMED LOAD FACTOR:	0%
BURN TIME:	0
ASSUMED NORMAL "OFF" TIME:	0%
NET HOURS:	0
INCREMENTAL COST/KWH:	$0.0000

OPERATING ASSUMPTIONS:

00% HOURS @ 100% FLOW
00% HOURS @ 0% FLOW
00% HOURS @ 0% FLOW
00% HOURS @ 0% FLOW

ENERGY CALCULATIONS:

ACTIVE KW:	0.0
KWH REDUCTION:	0
ENERGY SAVINGS:	$0
PAYBACK PERIOD:	ERR

COST ASSUMPTIONS:

QTY:	0
HP:	
COST:	$0
REBATE:	$0
NET COST:	$0

SYSTEM ASSUMPTIONS:

EQUIV. LIN. FT.	0
STATIC HEAD; FT.	0 ***(a)***
AVG. PIPE DIA. (IN.)	0
DESIGN GPM:	0 ***(b)***
PUMP EFFICIENCY	00%
MOTOR EFF.	00%

(a) If there are no plans or specifications to get this from, make reasonable assumptions.

(b) Same as above. Or use the equation given before: $GPM = 15 * D^2$

Be creative and play with all the various values in the "Systems Assumptions" that are not specifically known so you can observe the effect on the maximum horsepower and TDH. You will begin to see the effects of variations on the system, and, quite often, it will point out how oversized a pump really is.

WATER PUMP LOAD DATA: CIRCULATING PUMP IDENTIFICATION

	(n) GPM	*(o)* FT. TDH	*(p)* % SPEED	*(q)* MOTOR HP	*(r)* PSI TDH	*(s)* MOTOR KW
(1)	3.4	288.69	97%	4.76	125	4.1
(2)	6.7	288.90	97%	4.76	125	4.1
—	—	—	—	—	—	—
—	—	—	—	—	—	—
(14)	46.9	302.67	100%	5.11	131	4.4
(15)	50.3	304.74	100%	5.16	132	4.4

(n) *(b) is inserted into the last row then divided into 15 ascending steps (arbitrarily)*

(o) $\{[0.001246 * (n)^2 * \text{Equiv. Lineal Feet}] \div (\text{Pipe Diameter in inches})^5\} + \text{Static Head}$
See "Standard Plumbing Engineering Design"; Pg. 142 – reference above equation

(p) $(p)_x = \sqrt{\frac{(a)}{[\text{Col. }(o)_{15}]}} + \left[\frac{(n)_x}{(n)_{15}} * \left(1 - \sqrt{\frac{(a)}{[\text{Col. }(o)_{15}]}}\right)\right]$

(q)$_{15}$ $(\text{GPM} * \text{TDH}) \div (3960 * \text{Pump Efficiency}) = ((n)_{15} * (o)_{15}) \div (3960 * \text{Pump Eff.})$

(q)$_x$ $\text{Percent Speed}^3 * HP_{max} = (p)_x^3 * (q)_{15}$

(r)$_x$ $TDH_x * 0.4331 = (o)_x * 0.4331$

(s)$_x$ $(\text{Motor } HP_x * 0.746) \div \text{Motor Efficiency} = ((q)_x * 0.746) \div \text{Motor Efficiency}$

For pumps that are used in comfort cooling applications, utilize the local area temperature history through its bin temperature observation data. First you must make an assumption of the percentage of the building cooling load that is caused by the internal heat gains (people, equipment, etc.) and the remaining percentage is assumed caused by the building envelope heat gains. By assuming that the highest outside temperature represents 100% load, the lower temperatures represent a lower percentage load and thus a corresponding lower flow requirement. This percentage flow requirement would then be cubed when applying the Affinity Laws.

$$\% \text{ Load at Bin Temp.} =$$

$$(\% \text{ Internal load influence}) + \left\{ \left[\frac{(Bin\ Temp. - Design\ IAT)}{(Max.\ Bin\ Temp. - Design\ IAT)} \right] * (\%\ Building\ envelope\ influence) \right\}$$

This equation assumes that the system was designed reasonably correct, i.e., that the highest outside temperature applies the maximum load: 100%. The temperature bins below the highest bin represent a decreased load due to a lower Δt applied to the building envelope. The total load percentage is the internal loads percentage plus the reduced external loads percentage.

The following spreadsheet offers an approach that can be used for a chilled water circulating loop based upon the local temperature history.

CIRCULATING PUMP VARIABLE SPEED CONTROLLERS
CHILLED WATER CIRCULATING LOOP

BASE ASSUMPTIONS:

ACTIVE PUMP HORSEPOWER:	0
ASSUMED EFFICIENCY:	00%
ASSUMED LOAD FACTOR:	00%
BURN TIME:	0
ASSUMED NORMAL "OFF" TIME:	0%
NET HOURS:	0
ASSUMED BASELOAD SPEED:	00%
% HOURS @ BASELOAD SPEED:	00%
NET HOURS @ BASELOAD SPEED:	0
INCREMENTAL COST/KWH:	$0.0000

ENERGY CALCULATIONS:		COST ASSUMPTIONS:	
ACTIVE KW:	0.0	QTY:	0
KWH REDUCTION:	0	HP:	
ENERGY SAVINGS:	$0	COST:	$0
PAYBACK PERIOD:	ERR	REBATE:	$0
		NET COST:	$0

CHILLED WATER LOOP DATA

BASELOAD ASSUMPTIONS:		
	BUILDING ENVELOPE INFLUENCE:	00%
	INTERNAL LOADS INFLUENCE:	00%
	ANNUAL COOLING SEASON HOURS:	0

COOLING LOAD DATA
{CLOSEST CITY}

START MONTH TO END MONTH
HOURS PER DAY

DESIGN IAT: 00

BIN TEMP	BIN HOURS	ACCUM. BIN HOURS	% LOAD	ANNUAL % TIME	VSD KWH
97.5	7	7	100%	0.19%	0
92.5	34	41	93%	0.94%	0
87.5	122	163	86%	3.39%	0
82.5	249	412	79%	6.92%	0
77.5	396	808	73%	11.00%	0

% ANNUAL HOURS COOLING: 00%
% ANNUAL HOURS @ BASELOAD: 00%

(Example and calculations)

CIRCULATING PUMP VARIABLE SPEED CONTROLLERS
CHILLED WATER CIRCULATING LOOP

BASE ASSUMPTIONS:

ACTIVE PUMP HORSEPOWER:		15
MOTOR EFFICIENCY:		89%
ASSUMED LOAD FACTOR:		80%
RUN TIME:	*(a)*	3600
ASSUMED NORMAL "OFF" TIME:	*(b)*	0%
NET HOURS:	*(c)*	3600
ASSUMED BASELOAD SPEED:	*(d)*	65%
% HOURS @ BASELOAD SPEED:	*(e)*	78%
NET HOURS @ BASELOAD SPEED:	*(f)*	2792
INCREMENTAL COST/KWH:		$0.0675

(a) *Annual hours that the chilled water system is "enabled"*

(b) *Assumed % that controls may be satisfied causing the system to shut off; usually 0%*

(c) *(b) * (a)*

(d) *Automatically input from "Internal Loads Influence" cell*

(e) *Automatically input from bin data table; "% Hours At Base Load"*

(f) *(e) * (c)*

ENERGY CALCULATIONS:

ACTIVE KW:	*(g)*	10.1
KWH REDUCTION:	*(h)*	24,588
ENERGY SAVINGS:	*(i)*	$1,660
PAYBACK PERIOD:		3.37

COST ASSUMPTIONS:

QTY:	1
HP:	15
COST:	$7,590
REBATE:	$2,000
NET COST:	$5,590

(g) *(Active HP * Load Factor * 0.746) ÷ Motor Efficiency*

(e) $[(g) * (c)] - [(\text{Sum of Col. } (o)) + ((g) * (d)^3 * (f))]$
{existing kWh} – [{kWh under VSD control} + {kWh @ minimum, baseload speed}]

(i) *(h) * Incremental $ / kWh*

CHILLED WATER LOOP DATA

BASELOAD ASSUMPTIONS:			
	BUILDING ENVELOPE INFLUENCE:	***(k)***	35%
	INTERNAL LOADS INFLUENCE:	***(l)***	65%
	ANNUAL COOLING SEASON HOURS:	***(a)***	3600

(k) *Assumed percentage of the cooling load caused by heat gains through the building envelope and from infiltration, etc. This is the part of the load that responds to bin temperatures*

(l) $(1-(k))$ *Assumed percentage of the cooling load caused by heat gains from occupants and equipment. This is the part of the load that represents the base load on the system.*

COOLING LOAD DATA

{ALBANY, NY}

MAY 1 TO SEPTEMBER 30
24 HOURS PER DAY

DESIGN IAT: 72

BIN TEMP	BIN HOURS	ACCUM. BIN HOURS	*(m)* % LOAD	*(n)* ANNUAL % TIME	*(o)* VSD KWH
97.5	7	7	100%	0.19%	70
92.5	34	41	93%	0.94%	276
87.5	122	163	86%	3.39%	788
82.5	249	412	79%	6.92%	1,254
77.5	396	808	73%	11.00%	1,521

% ANNUAL HOURS COOLING: 22.4% ***(p)***
% ANNUAL HOURS @ BASELOAD: 77.6% ***(q)***

(m) $(l) + \left(\left(\left(\text{Bin Temp.} - \text{Design IAT}\right) \div \left(\text{Max. Bin Temp.} - \text{Design IAT}\right)\right) * (k)\right)$
Assumes that the system was designed reasonably correct, i.e., that the highest temperature applies the maximum load: 100%. The temperature bins below the highest bin represent a decreased load due to a lower Δt applied to the building envelope. The total load percentage is the internal loads percentage plus the reduced external loads percentage.

(n) $\left(\text{Bin Hours} \div (a)\right) * 100$

(o) $(m)^3 * \text{Bin Hours} * (g)$ *{At last! The application of the Affinity Law.}*

(p) $\left(\text{Total Accumulated Bin Hours} \div (a)\right) * 100$

(q) $\left(100 - (p)\right)$

Evaluating the energy savings for boiler feed water pumps, basically the only application for a VSD exists when the pump runs continuously against a modulating valve that controls the flow. Because the static head is usually high (150 to 250 feet) the minimum speed to overcome the boiler pressure is usually higher than 90%.

So where are the savings? Even 95% cubed results in a 14% savings. Since these are typically long run time situations (all winter for heating or all year for process applications) the savings are often sufficient to justify the implementation. In addition, very often the pumps are so oversized that the flow control valve is wasting a lot of energy by throttling, and the output pressure of the pump is very much higher than needed.

There *are* reasons for over sizing boiler feed water pumps. First, having each pump sized to serve all the boilers in a multiple boiler installation adds redundancy. Second, if the pump is sized only to handle the equivalent gpm of the steam flow it will never be able to raise the boiler water level during periods of peak usage. Finally, if it can only just keep up, it will not be able to handle any emergency requirement to raise the water level in the boiler in a hurry. This is one application where I don't recommend downsizing. If you don't change the ability of the pump to deliver whatever peak flow it was designed to produce, you can't be sued for screwing things up. (*Successfully* sued, that is. You can *always* be sued.)

Typically when you apply the water horsepower formula you will be amazed when you calculate a required horsepower 1/2 to 1/3 of the connected horsepower. This is the horsepower that will be delivered by a VSD.

The following spreadsheet offers an approach to calculating the savings with a boiler feedwater pump.

BOILER FEED WATER PUMP

BASE ASSUMPTIONS:	
ACTIVE PUMP HORSEPOWER:	0
ASSUMED EFFICIENCY:	0%
ASSUMED LOAD FACTOR:	0%
BURN TIME:	0
ASSUMED NORMAL "OFF" TIME:	0%
NET HOURS:	0
INCREMENTAL COST/KWH:	$0.0000

OPERATING ASSUMPTIONS:				
00%	HOURS	@	100%	FLOW
00%	HOURS	@	0%	FLOW
00%	HOURS	@	0%	FLOW
00%	HOURS	@	0%	FLOW

ENERGY CALCULATIONS:	
ACTIVE KW:	0.0
KWH REDUCTION:	0
ENERGY SAVINGS:	$0
PAYBACK PERIOD:	ERR

COST ASSUMPTIONS:	
QTY:	0
HP:	
COST:	$0
REBATE:	$0
NET COST:	$0

SYSTEM ASSUMPTIONS:	
EQUIV. LIN. FT.	0
STATIC HEAD; FT.	0
BOILER PRESSURE (PSIG):	0
AVG. PIPE DIA. (IN.)	0
BOILER HP	0
PPH; STEAM	0
GPM RETURN RATE	0
GPM @ 2.5 SF:	0
PUMP EFFICIENCY	00%
MOTOR EFF.	00%

WATER PUMP LOAD DATA: BOILER FEED WATER PUMP

GPM	FT. TDH	% SPEED	MOTOR HP	PSI TDH	MOTOR KW
0	0	0	0	0	0
0	0	0	0	0	0
0	0	0	0	0	0
0	0	0	0	0	0
0	0	0	0	0	0
0	0	0	0	0	0
0	0	0	0	0	0
0	0	0	0	0	0
0	0	0	0	0	0
0	0	0	0	0	0
0	0	0	0	0	0
0	0	0	0	0	0
0	0	0	0	0	0
0	0	0	0	0	0
0	0	0	0	0	0

(Example and calculations)

CIRCULATING PUMP VARIABLE SPEED CONTROLLERS
BOILER FEED WATER PUMP

BASE ASSUMPTIONS:

ACTIVE PUMP HORSEPOWER:		10
ASSUMED EFFICIENCY:		87%
ASSUMED LOAD FACTOR:		80%
RUN TIME:	*(a)*	8760
ASSUMED NORMAL "OFF" TIME:	*(b)*	0%
NET HOURS:	*(c)*	8760
INCREMENTAL COST/KWH:		$0.0000

OPERATING ASSUMPTIONS:

	(d)		*(e)*	
(1)	25%	HRS @	100%	FLOW
(2)	25%	HRS @	75%	FLOW
(3)	25%	HRS @	50%	FLOW
(4)	25%	HRS @	25%	FLOW

(a) Annual hours that the boiler is active.
(b) Assumed % that controls may be satisfied causing the system to shut off; usually 0%
(c) (b) * (a)
(d) Operating assumptions defining the amount of time that each flow step will experience
(e) Operating assumptions defining the variations in flow in discrete steps

ENERGY CALCULATIONS:			COST ASSUMPTIONS:	
ACTIVE KW:	(f)	6.9	QTY:	1
KWH REDUCTION:	(g)	23,017	HP:	10
ENERGY SAVINGS:	(h)	$1554	COST:	$7,245
PAYBACK PERIOD:		3.5	REBATE:	$1,800
			NET COST:	$5,445

(f) *(Active HP * Load Factor * 0.746)* ÷ Motor Efficiency

{Existing kWh} – {Step 1 kWh – Step 2 kWh – Step 3 kWh – Step 4 kWh}

(g) *((f)*(c)) – [(d1)*(c)*@VLOOKUP(((m)*(E1)), "Load Data Table", (s))] –*
[(d2)(c)*@VLOOKUP(((m)*(E2)), "Load Data Table", (s))] –*
[(d3)(c)*@VLOOKUP(((m)*(E3)), "Load Data Table", (s))] –*
[(d2)(c)*@VLOOKUP(((m)*(E4)), "Load Data Table", (s))] –*

(h) *(g) * Incremental $/kWh*

SYSTEM ASSUMPTIONS:		
EQUIV. LIN. FT.	500	
STATIC HEAD; FT.	289	(t)
BOILER PRESSURE (PSIG):	125	(u)
AVG. PIPE DIA. (IN.)	2.5	
BOILER HP	300	(i)
PPH; STEAM	10,056	(k)
GPM RETURN RATE	20	(l)
GPM @2.5 SF:	50	(m)
PUMP EFFICIENCY	75%	
MOTOR EFF.	87%	

(u) *Manual Data Entry*
(t) *(u) ÷ 0.4331*
(i) *Manual Data Entry; or: PPH ÷ 33.52*
(k) *(i) * 33.52*
(l) *(k) ÷ 500*
(m) *(l) * 2.5*

WATER PUMP LOAD DATA:BOILER FEED WATER PUMP

	(n) GPM	(o) FT. TDH	(p) % SPEED	(q) MOTOR HP	(r) PSI TDH	(s) MOTOR KW
(1)	3.4	288.69	97%	4.76	125	4.1
(2)	6.7	288.90	97%	4.76	125	4.1
—	—	—	—	—	—	—
—	—	—	—	—	—	—
(14)	46.9	302.67	100%	5.11	131	4.4
(15)	50.3	304.74	100%	5.16	132	4.4

(n) *(m) is inserted into the last row then divided into 15 ascending steps (arbitrarily)*

(o) $\{[0.001246 * (n)^2 * \text{Equiv. Lineal Feet}] \div (\text{Pipe Diameter in inches})^5\} + \text{Static Head}$
See "Standard Plumbing Engineering Design"; Pg. 142 – reference above equation

(p) $$(p)_x = \sqrt{\frac{(t)}{[Col.\ (o)_{15}]}} + \left[\frac{(n)_x}{(n)_{15}} * \left(1 - \sqrt{\frac{(t)}{[Col.\ (o)_{15}]}}\right)\right]$$

$(q)_{15}$ $(GPM * TDH) \div (3960 * \text{Pump Efficiency}) = ((n)_{15} * (o)_{15}) \div (3960 * \text{Pump Eff.})$

$(q)_x$ $\text{Percent Speed}^3 * HP_{max} = (p)_x^3 * (q)_{15}$

$(r)_x$ $TDH_x * 0.4331 = (o)_x * 0.4331$

$(s)_x$ $(\text{Motor } HP_x * 0.746) \div \text{Motor Efficiency} = ((q)_x * 0.746) \div \text{Motor Efficiency}$

ANECDOTES

Richard's Retrofit Rules

Pumping Rule #1: *"You can never save more energy than shutting it off."*

Pumping Rule #2: "You will never get a production manager to accept downsizing."

Pumping Rule #3: *"No one in production knows the flow requirements."*

The very first VSD installation I designed involved a 50 horsepower pump in a textile dyeing application. This pump was attached to a 6" diameter pipe with two short elbows immediately connected to the outlet that fed into a pneumatic butterfly valve. The operator made sure that the valve was closed during start up, then he would gradually open the valve until the flow in the open tank was visually satisfying to him. The pump would then run for two to four hours at a time with fifteen minute breaks while the quality was checked. The total daily run time was close to 22 hours.

Because I could visually see that the butterfly valve was only open a fraction I knew that there was potential here. I installed pressure gages on both sides of the valve. The pressure on the inlet side of the valve was 60 psig and the pressure on the outlet side of the valve was only 5 psig.

This confirmed that there was a major energy loss across that valve. In order to quantify the savings I needed to find out what the flow rate was. See rule #3. Not even the dye house Chemist who is responsible for designing the chemical formulas knew how much water was going through the fabric. Fortunately, the equipment manufacturer was willing to make a guess at 750 gpm. The chemist thought that sounded OK, so that's what we used.

Running through the standard pump formulas, I calculated that the required pump brake horsepower for that system without a valve was *five horsepower*! My first reaction was that I made some stupid mistake. Since I didn't want anyone to know that I was stupid I spent the next day studying the application and pump formulas in general. I couldn't get away from the five horsepower answer.

Finally, I called the equipment manufacturer and asked them where I was going wrong. Here *I* was saying that there is only a five horsepower load and there *they* were using 50 horsepower on all their equipment. They were very cooperative and reassuring. They agreed that five horsepower probably was all that was needed in our application of their equipment, but they sell their equipment to all kinds of dyeing operations. Some of the other operations will need that 50 horsepower. Since you are paying for it, they keep their design and inventory standardized at 50 horsepower.

Armed with new self-confidence, I recommended that the 50 horsepower motors be removed and replaced with 10 horsepower motors. (Why go *too* far out on a limb when you don't have to?) Then we can install the VSD equipment at the much cheaper 10 horsepower level.

Not a chance.

The production supervisors got sweaty palms and the production manager broke out in a rash. "But we may *need* that 50 horsepower some day!"

Although my conversations with the equipment manufacturer outlining the different dyeing operations that would require 50 horsepower had led me to conclude that we would never, never, *never* need 50 horsepower, I explained that the motors could easily be reinstalled when the time called for it. And that the difference in the cost to install a 10 horsepower VSD vs. a 50 horsepower VSD was about $6,000 for each one of the five pumps.

Money was no object when it came to production reliability, translated: their peace of mind. The 50 horsepower units were installed. They still saved a lot of money. They even improved quality with better control of the flow and pressure. But the cost to implement could have been much cheaper.

One thing I'd like to caution against: resist the temptation to install one VSD on a system with two or more pumps discharging into the same header. The seductive idea is to run the VSD pump until the flow demand is greater than one pump's flow, but less than both. The theory is that you can run one pump at full speed and the other one at increasing speeds to meet the increasing flow requirements up to two full pumps. This sounds like a good way to get the benefit without buying two systems. Believe me, if I thought it would work I would be recommending it loud and long.

Unfortunately, it doesn't pass the Affinity Law test.

Imagine that each pump has a check valve installed at its outlet. If one pump is running at full speed, it will be putting out full pressure into the header. If the second pump is running at, say, 50% speed to give a total system flow of 150%, the outlet pressure must be 0.50^2, or 25%. Obviously, if the pressure from one pump is four times higher than the pressure from the other pump, one check valve will be open but the other will be closed.

These systems *will* function and may give the impression that all is well. To put is simply, the feedback signal controlling the speed of the second, variable speed pump, whether it is sensing flow or pressure, cannot be satisfied until the second pump has an impact on the system. This will cause the feedback signal to keep calling for increased speed until the discharge pressures match and the "check valve" opens. The flow to the system will be satisfied, but the energy savings will be negligible, if any.

Chapter 4

Fans

INTRODUCTION

The evaluation of fan conservation measures is very similar to pump measures. This is so because most fans are centrifugal pumps that pump air. Therefore the affinity laws apply the same way as for pumps.

The differences lie in the fact that most fan systems do not have a minimum static head.

Many building exhaust fan systems are grossly oversized and over-operated. They are designed, rightly so, for the highest occupancy conditions and run that way continuously for 24 hours a day. A VSD will allow plant personnel to match the exhaust quantities to the exhaust requirements without making major modifications.

The best part of this is that you can reduce the exhaust by duty cycling without completely stopping it. A duty cycle that runs the exhaust fan at 50% speed, therefore 50% flow, for five minutes out of every fifteen minutes will reduce the horsepower per hour to 71%. The amount of exhausted air will be reduced by 17% thereby reducing the load on the HVAC system. This will not add any excess strain or wear on the equipment and, more importantly, no one will notice.

Cooling tower fans provide additional air circulation through cascading water to increase the cooling effect of conduction, convection and evaporation. As the air temperature drops (specifically the wet bulb) less air flow will be required to provide the same cooling effect. A VSD controlled by discharge water temperature will allow the fan speed and energy to automatically match the cooling requirements. (If the cooling tower pump VSD application is also implemented, the controls should be designed such that the pump is brought up to full speed before the fan is enabled to start at slow speed.)

Many variable air volume (VAV) systems were originally designed without cfm control devices. Second generation VAV systems utilized pressure relief and bypass dampers. Third generation systems utilize variable inlet vanes to control fan output. All these systems can benefit from the installation of a VSD and removal of the other flow control methods. Specifically, the conversion of a variable inlet vane flow control system to a VSD flow control system is usually very simple to implement. In most cases the existing flow control sensors and feedback hardware are directly compatible with the controls for the VSD.

Induced and forced draft fans for larger boilers, rotary kilns, process furnaces, etc. generally control the flow of air through vanes or dampers to control the amount of combustion air as needed to match the firing rate of the equipment. Whether inlet vanes or discharge dampers are used, significant savings can be realized if the flow is controlled by a VSD. In many cases, especially the larger pieces of equipment, the discharge pressure is relatively high due to large static pressure losses through the firing chamber, heat exchanger, and various pollution control devices. This will limit the minimum speed to whatever is necessary to overcome those pressure requirements.

THINGS TO LOOK OUT FOR ARE:

1. *Dust collection system fans.* These systems require a minimum cfm to maintain carrying velocity throughout the duct system. Long duct lengths with high flow volumes require a relatively high static pressure. High static pressure fans often are not centrifugal fans. Certain fan speeds may result in resonant frequency vibration amplification (this can be avoided with VSD options).

2. *Fume Hood exhaust fans.* Generally, there is little variation allowed in a single fan and hood system. Variations are possible in a laboratory with multiple fume hoods served by a single fan. However, the system is there for health reasons. I maintain a personal policy that life safety and health systems are not worth the liability to change them for energy savings only. If there is a need for a new design or redesign then energy efficiency can be designed into the system with the help and guid-

ance of the health safety experts that set the requirements.

DEFINITIONS

{As you will see, most of these terms are repeated in different sections. I have chosen to repeat them because I felt that each section should be able to stand on its own as a reference. (Not because I needed to fill space in the book.)}

Affinity Law

I had to look up the word "affinity" in the dictionary to get an idea why it was chosen for the formulas defining the relationships between speed ratios, head ratios and horsepower ratios for centrifugal pump and fan applications. I found out that the word "affinity" means "relationship". It is my feeling that a deliberate effort is made every now and then to name useful and clever formulas with words that tend to hide the connection to their use in the real world. Possibly it is just a result of engineers taking a cue from doctors and lawyers.

Anyway, the ratio equations are straight forward and simple to manipulate to determine any useful information with a few known facts:

$$\frac{N_1}{N_2} = \frac{F_1}{F_2}; \quad \left(\frac{N_1}{N_2}\right)^2 = \frac{H_1}{H_2}; \quad \left(\frac{N_1}{N_2}\right)^3 = \frac{HP_1}{HP_2}$$

N = rpm F = flow H = head HP = horsepower

Pressure

With respect to air flow, this term is synonymous with the word "Head". By far in air flow situations it most often refers to pressure given in units of Inches of Water Column, other times it may be Pounds per Square Inch.

Static Pressure

The static pressure is the pressure at the fan that is developed or required even when there is zero flow in the system (static conditions).

Velocity Pressure

The velocity pressure is the pressure losses due solely to the friction of the air moving against the ductwork and fittings.

Pressure Drop (PD)

The pressure drop is the change in pressure between the inlet of something and the outlet of something. In an air handling system, the pressure drop from the beginning to the end is the velocity pressure.

All pieces of equipment will specify what the pressure drop through the equipment is expected to be when you operate it at its rated capacity. This is the friction head that needs to be overcome in order to push air through the equipment at the flow rate for which it is designed.

If the equipment is in parallel, the equipment with the highest PD is the fan controlling factor. If the equipment is in series, the sum of each PD is the fan controlling factor.

Total Pressure (TP)

The TP is the highest pressure that the ductwork and equipment system will impose on the fan. This is simply a sum of the static pressure and velocity pressure.

CALCULATIONS:

Remember, you have two goals:

1. Correctly define the existing conditions in terms of total energy consumption
2. Estimate the total energy consumption that will occur *after* the project is implemented.

The first goal is simply the HP of the fan motor times the hours it runs.

The second goal is simply the HP of the fan motor at reduced speeds (flows) times the hours that it runs at those reduced speeds.

To find the horsepower at *any* flow you simply have to fill in the unknowns in the following equation:

$$\text{Horsepower} = \frac{\text{CFM} * \text{Pressure (In. W.C.)}}{6356 * \text{Fan Eff.}}$$

Notice the similarity to the water horsepower equation. Basically, they are both the flow times the pressure, divided by efficiency and a constant for units conversion.

Determining the various flow rates (cfm) that are required to meet the system's functional needs is what you get paid to do. If you didn't think that there would be any variations then you wouldn't be evaluating a VSD.

However, applying the Affinity Law ratios will simplify the calculation when the horsepower and flow at full speed is known.

$$\frac{N_1}{N_2} = \frac{F_1}{F_2}; \quad \left(\frac{N_1}{N_2}\right)^2 = \frac{P_1}{P_2}; \quad \left(\frac{N_1}{N_2}\right)^3 = \frac{HP_1}{HP_2}$$

N = rpm F = flow P = pressure HP = horsepower

$$\left(\frac{F_1}{F_D}\right)^3 * HP_D = HP_1$$

F = flow HP = horsepower

There are really only a few applications that I have run into that require the inclusion of the minimum speed formula. However, one of them comes up rather a lot.

The feedback controls for the fan speed of a variable air volume system operate on the concept of maintaining a constant duct pressure at a specific design point in the duct system (rule of thumb is to place the sensor 2/3 of the way down stream from the fan outlet). At first this may sound like a no win situation because if the pressure remains constant then the pressure ratio equals one. Therefore the speed ratio must also equal one (the square root of one). If the speed cannot change for the system to work, then there cannot be any horsepower savings.

Where are the savings? Remember that the pressure must remain constant 2/3 of the distance down stream from the fan outlet. The fan must present a discharge pressure that is higher than the set point pressure in order to overcome friction losses along the way. In addition, branches and zone discharges both upstream and down stream will affect the pressure at the set point. Therefore the fan must increase and decrease its discharge pressure as the flow varies in order to maintain the set point down stream.

But there *is* a minimum pressure that the fan must produce at zero flow to produce the pressure at the set point. This will result in the need for a minimum speed and horsepower requirement just as calculated for a pump.

$$(\%\ Speed)_{Min} = \sqrt{\frac{H_{MIN}}{H_D}}$$

The H_{Min} in this situation is the pressure set point, based on the assumption that with zero flow the pressure down stream will be equal to the fan discharge pressure. The H_D is the set point pressure plus the friction losses at design flow. Hopefully, since VAV systems are relatively new, the design conditions will be easily available.

ANECDOTES

Richard's Retrofit Rules

Fan Rule #1: *"You can never save any more energy than shutting it off. But when you do, someone will complain that it's too ________ (stuffy, cold, hot, quiet, etc.)."*

Fan Rule #2: *"It is always a surprise to find out that fan noise is more important than the actual movement of air."*

Fan Rule #3: *"Employees always believe that more exhaust is better."*

While performing an energy audit at a facility that made light aggregates for cinder blocks by heating shale in a rotary kiln that was 12 feet in diameter and about 150 feet long, I observed that they had already installed a VSD on a 400 horsepower forced draft fan. This eliminated the use of a throttling damper to control the firing rate of the kiln. While I was there a second kiln was down for repairs and upgrade. Facility personnel reported that the energy savings calculations may have justified the first installation, but they hadn't bothered to make actual measurements because their bill went down and process control was improved. Therefore they had ordered a second VSD for the second kiln.

Chapter 5

High Efficiency Motors

INTRODUCTION

Replacing standard efficiency motors with high efficiency motors has become as common place as replacing incandescent fixtures with fluorescent. Most utility companies offer very attractive incentives. When performing a utility subsidized energy audit most utilities even require that an energy auditor must include such an evaluation in their reports. The reality is that quite often the predicted savings do not occur.

I happen to believe that it is economically smart to implement a policy to use high efficiency motors as replacements and specify them for new equipment. However, it is rare to find a case where replacement of a working motor is economically sound. The contemporary wisdom defining how to evaluate the savings consistently predicts higher savings than actually is achieved. As it turns out, the reasons for this failure are not that difficult to understand.

DEFINITIONS

High Efficiency

I figured that I would start with the hardest first. Or maybe this is the easiest. Since every manufacturer has their own definition of "high efficiency" it generally means "something higher than our standard efficiency motor".

Certainly, NEMA has come out with standards (or should I say; ever changing standards) that define minimum efficiencies at *full load horsepower* to qualify as high efficiency. These are dutifully stamped on the nameplates of all high efficiency motors. But the best you can say is that if you ever operate that motor at full load, it should work that efficiently.

Next come the utility companies. They usually set their own minimum standard efficiency for their incentive programs. These are quite often higher standards than NEMA. Therefore they usually call them "Premium Efficiency" standards.

Synchronous Speed

The synchronous speed is the ideal maximum rpm attainable with the number of magnetic poles in the motor windings and the frequency of the power supply. The higher the number of poles, the lower the synchronous rpm. The synchronous speed formula is:

$$\frac{Hertz * 120}{\# \text{ of Poles}}$$

Slip

Slip is the difference between the synchronous rpm and the actual rpm of a motor. The higher the efficiency, the lower the slip.

WHAT TO LOOK FOR

Any motor can be a candidate for replacement with a higher efficiency model on burn out. The best indicator for economic viability is long run hours.

WHAT TO LOOK OUT FOR

Motors coupled directly to pumps and fans. These can certainly result in savings, but the savings will be considerably *less* than expected because the rpm will be slightly higher. In fact, under certain conditions, the energy consumed can be significantly *higher* than with the standard efficiency motor.

CALCULATIONS

I promised you simple solutions and this equation is the simplest.

To determine the theoretical savings from motor efficiency improvement the formula is as follows:

$$kW\ Saved = 0.746 * hp * \left[\left(\frac{1}{Eff.\ Motor_1}\right) - \left(\frac{1}{Eff.\ Motor_2}\right)\right]$$

One important thing to remember is that the horsepower in this equation should be the actual horsepower load on the motor. The equation defining the real, not the theoretical, savings includes the percent load on the motor. Otherwise known as the load factor:

$$kW\ Saved = 0.746 * hp * LF * \left[\left(\frac{1}{Eff.\ Motor_1}\right) - \left(\frac{1}{Eff.\ Motor_2}\right)\right]$$

Therein lies this measure's "Catch 22". If the horsepower load on the motor is less than 50% of its nameplate then all bets are off as to the motor efficiency. If the motor is loaded up to 100% then it should be evaluated to be sure it is all right to operate that way for long periods. Otherwise it may be smart to install a larger motor.

What you need in most cases is the value of the "part load" efficiency which is usually given for 25%, 50%, and 75% loading. That is, when you can get such information.

The second problem arises when the motor is direct-coupled to a centrifugal load. If you remember the Affinity Law from previous chapters, you learned how a small reduction in rpm can have a major effect on the horsepower. Well, not surprisingly, it applies to increases in motor rpm as well. The increased motor efficiency necessarily results in a reduction in the motor slip. Therefore the rpm increases slightly as the efficiency increases.

An increase of 10 rpm on a motor that normally ran at 1750 rpm means the speed ratio of before and after will be 1.006:

$$\frac{1760}{1750} = 1.006$$

Which may not seem significant. However, if the motor is driving a centrifugal load, then the Affinity Law causes the horsepower to increase to 102%.

$$\left(\frac{1760}{1750}\right)^3 = 1.006^3 = 1.02$$

When the high efficiency motor is expected to save only 3% to 5% of the full load horsepower, adding another 2% to the load can eat away more than half of the savings.

The theoretical savings formula should be modified as follows to reflect the differences in horsepower before and after.

$$kW\ Saved = 0.746 * hp_1 * \left(\frac{1}{Eff.\ Motor_1}\right) - 0.746 * hp_2 * \left(\frac{1}{Eff.\ Motor_2}\right)$$

The distinguished engineer Mr. Konstantin Lobodovsky has shown that in certain applications, the conversion to high efficiency motors actually uses *more* energy than the previous motor in that application.

Finally, once you have the correct kW reduction, all you have to do is multiply it by the annual hours that the motor runs to get the total kWh reduction.

ANECDOTES

Richard's Retrofit Rules

> **Motor Rule #1:** *"You can never save any more energy than shutting it off."*
>
> **Motor Rule #2:** *"No one in the plant knows how much the motor is loaded."*
>
> **Motor Rule #3:** *"If you burn your hand when you touch it, it is at least 100% loaded."*

Being the project manager of a large motor retrofit job I was expected to have all the answers. With the risk of sounding defensive, the motor evaluation was required by the local utility as part of a major energy conservation report and the evaluation was performed by someone else utilizing the standard savings formula.

As work progressed, the very savvy operations and maintenance personnel asked some hard questions. Questions like:

"If this is supposed to save energy, why are my motor ampere readings higher than with the standard motor?"

"If this is supposed to save energy, why are the motor temperature readings higher than before? And isn't the higher temperature going to adversely effect the life of the motor?"

Of course I was afraid to answer those questions the way that I really felt. The utility company would have choked and then my boss would have choked....me. So I deferred to the motor manufacturer's representative. The end result was that some of the motors were removed as defective and replaced. In most cases the replacement motors produced amp readings lower than before, if only slightly. Other motors were replaced with heavier duty motors to lower operating temperature. The representative also said that the higher temperatures were the result of better heat transfer characteristics in the design of the motor housing that allow the cooling fan to be downsized, thus reducing the energy losses called "windage" and would not be detrimental to these motors.

Ultimately, the customer was satisfied because he managed to get all new motors for a very low net cost after the utility rebate. I was satisfied because ultimately I didn't have to answer any of those questions.

However, the bottom line should be emphasized. The majority of the motors at that site were not direct-coupled. The specifications called for tachometer readings before replacement, and sheave adjustments to maintain the same output rpm. Therefore the majority of the motors showed lower amperes and temperatures.

Chapter 6

Insulation

INTRODUCTION

The evaluation of conservation measures involving insulation starts with your ability to feel warmth and cold. Honestly, some of the best measures that I have identified in practice came to me when I came close to burning myself. It's pretty safe to say that there *will* be some savings potential if you can feel the heat even before you touch it.

Pipes are easy. Especially hot pipes. The evaluation relies upon determining the heat loss from conduction, convection, and radiation. At first this may seem complicated. If you were required to determine the various heat transfer factors and coefficients, etc. ad infinitum ad nauseam it would be *very* complicated. But would I be talking about this if I didn't have an easy way out? Fortunately for me, Owens Corning, Manville, Dow, and everyone else trying to sell insulation have done all the work. They all publish tables showing the total heat loss from various diameter pipes. You get the heat loss in Btuh per lineal foot for horizontal and vertical orientations, with wind, without wind, at all temperatures, with weatherproof jackets, without jackets, and most important – bare pipe.

Now here's the part where I cheat:

If I have observed a 30" diameter condensate return tank that is bare steel holding 200° F water, I don't have to perform the heat loss calculations for a bare cylinder if I have a table that gives me the values for a 30" diameter pipe. Honestly, the various laws of physics really can't tell whether the object is a pipe or a tank. Yes, the pipe doesn't have any ends. Therefore your analysis is conservative at worst. If the payback calculation is only marginal then you can add the heat loss from the ends. My experience is that this conservative approach shows payback periods in months.

That is the essence of this book. The fast analysis that tells you whether it's worth going deeper.

When the surface is clearly not a pipe, even when you stand back and squint your eyes, then it must be a flat surface. Fear not! The same sources have developed heat loss tables for vertical and horizontal surfaces. There you can get the Btuh per square foot for the tank ends that were missing from the pipe heat loss table. Yes, most tank ends are curved, not flat. So what? Again the result is that your analysis is conservative.

DEFINITIONS

{As you will see, most of these terms are repeated in different sections. I have chosen to repeat them because I felt that each section should be able to stand on its own as a reference. (Not because I needed to fill space in the book.)}

Btu

This stands for British Thermal Unit. A Btu is defined as the amount of heat that it takes to increase the temperature of one pound of water by one degree Fahrenheit. Do not confuse Btu with temperature! The complete combustion of a standard wooden kitchen match will release approximately one Btu. The *temperature* of the flame will be over 1200° F. To get a perspective on quantities of Btus, one gallon of gasoline contains approximately 128,000 Btus.

Btuh

Simply means Btu per hour.

MBH

Simply means 1,000 Btu per hour.

mmBtu and mmBtuh

Means 1,000,000 Btu and Btuh.

Conduction

The transmission of heat by molecular vibrations from one part of a body to another part of the same body or to another body in physical contact with it, without appreciable displacement of the particles of the body.

The key idea here is physical contact.

R-factor; Thermal Resistance

Resistance is a property of a material that determines the amount of heat that will flow through a unit area given a difference in temperature.

The R-factor is the measure of the ability of a material to resist letting heat pass through it to a colder area and so it is otherwise called the thermal resistance. The higher the R-factor the better it is. The units of measure for R-factor are Δ°F/Btuh/sq.ft.

U-factor; Over-all Coefficient of Heat Transfer

Time rate of transfer of heat by conduction, across unit area for unit difference of temperature.

The U-factor is the measure of the ability of a material to let heat pass through it to a colder area and so it is otherwise called the thermal conductivity. It is the mathematical inverse of the sum of all the R-factors . The lower the U-factor the better it is. The units of measure for U-factor are Btuh/sq.ft./°F(Δt).

End-use efficiency

Sources of thermal energy generally have to deal with two areas where heat is lost before it arrives at the "end-use". There are combustion losses, i.e., heat lost up the stack. Then there are transmission losses, i.e., heat lost to the areas that the pipes or ducts pass through before they reach the end-use. The end-use efficiency is used to determine the amount of fuel that needs to be burned at the boiler in order for there to be the required amount of Btus at the end-use. The formula for calculating end-use efficiency is as follows:

*End-Use efficiency = Combustion efficiency * transmission efficiency*

Remember that heat lost by space heating pipes into the space that is to be heated is not really lost. All you lose is control.

CALCULATIONS

The best situation for cost effective insulation projects is when the heat source is electricity: the most expensive source of Btu's known to man. (Actually, the most expensive source of Btu's known to man so far is atomic energy. But we use that to make electricity, so by the time we get it, electricity is more expensive.) Electricity has the advantage of being 100% efficient at its end use. Therefore the conversion from units of Btu to units of kWh is very simple:

3,413 Btu = 1 kWh

Once you have determined the total Btu saved for the year:

*Btuh * Annual hours at the elevated temperature = Annual Btu loss*

You simply divide the annual Btu saved by 3,413 to get the annual kWh saved.

Other situations deal with fuel energy conservation. The heat lost by steam distribution pipes and fittings should not be overlooked. As a matter of fact, this loss must be calculated and steam traps sized correctly to transport the condensate away from the supply pipes or there will be serious consequences to the system.

To evaluate the heat loss, and cost of these heat losses, refer to the product literature as mentioned above. By subtracting the Btuh loss after the new insulation from the Btuh loss before the new insulation you will know how many Btuh will be saved.

Remember, the most important information that you are getting from these table is the heat loss for the bare surface. This is the information that is pure physics without any egocentric sales promises. The heat loss from their insulation may be subject to suspicion, but the loss from a bare surface is so high that once you become aware of it, to do nothing will be inexcusable (even for an accountant).

For example: the heat loss from the bare condensate tank described above is 1,857 Btuh per lineal foot. With 1.5" of fiberglass insulation with a bright metal jacket the heat loss drops to 198 Btuh per lineal foot. Therefore the Btuh saved is 1,659 Btuh per lineal foot. Assuming a five foot long tank that is hot for 3,600 hours during the winter, the annual Btu loss is 29,862,000 Btu per year.

How much will this save? The answer lies in the cost of the fuel and how long the system operates at elevated temperatures.

Suppose natural gas costs $5.00 per mmBtu. This is the price as it passed through the gas meter. The combustion efficiency should be no lower than 80% and the distribution efficiency should be around 85%. Therefore the end-use efficiency will be 68%:

$$End\text{-}Use\ efficiency = 0.8 * 0.85 = 0.68$$

This will be used to determine how many Btus had to be consumed at the burner in order to have those Btus to waste at the condensate tank.

$$(End\text{-}Use\ Btu) \div (End\text{-}Use\ Efficiency)$$

$$29{,}862{,}000 \div 0.68 = 43{,}924{,}706\ Btu$$

So the cost of the 1,659 Btuh per lineal foot that was lost before insulating the tank was:

$$\left(\frac{43{,}914{,}706}{1{,}000{,}000}\right) * \$5.00 = \$219.57\ Per\ year$$

The heat and cooling energy lost through the building envelope presents a very frustrating set of circumstances to me. First, it is usually a significant amount of energy and, more importantly, money. However, the money that would need to be spent to reduce the energy losses is typically 10 to 12 times greater than the savings. Unless, of course, you involve an architect in the project. Then there is no limit to the amount of money it will cost.

At any rate, you're welcome to do the analysis. The heat loss from conduction through a wall is calculated one of two ways:

$$Q = \left(\frac{T_o - T_i}{R}\right) * A;\ \&\ Q = U * A * (T_o - T_i)$$

Q = Btuh
T_o = Outside Air Temp.
T_i = Inside Air Temp.
A = Area; sq. ft.
R = R-factor
U = U-factor

When trying to calculate the Btuh saved with increased insulation, you can subtract the U-factors and multiply the answer times the area and the ΔT to get the savings. The trick is that you can't just subtract the R-factors. You must subtract the *inverse* of the R-factors.

$$\Delta Q = (\Delta T) * A * \left(\frac{1}{R_1} - \frac{1}{R_2}\right)\ \ \&\ \ \Delta Q = (U_1 - U_2) * A * \Delta T$$

ΔQ = Btuh saved
ΔT = (Outside Air Temp. – Inside Air Temp.)
A = Area; sq. ft.
R = R-factor
U = U-factor

I will talk briefly about infiltration. The first thing to say is that *nobody* can calculate the heat losses due to infiltration with any certainty. That is why the ASHRAE manual is so important. Because there is no way to determine the actual annual infiltration losses, when there is any dispute about the actual savings you can always point to the manual and say that you followed industry standards. There is always comfort in knowing that you used the most complicated method to arrive at an arbitrary answer.

So, since I believe that whatever answer I come up with is only a rough approximation, I use the simplest method that I know of to make my first cut. If the measure looks like it will be marginal but still desirable, I am forced to proceed with the ASHRAE methods to cover my tail.

The simplest method uses the total linear feet of perimeter around the windows and doors and applies a standard 0.25 cfm per foot to arrive at a total infiltration cfm. Then you simply apply the standard Btu equations for cfm to arrive at the heat loss. The trick in this method is to use only half of the total lineal feet of perimeter of all the windows. Why? Because the quantity of air that comes in must cause an equal quantity of air to be forced out. So while half the cracks are letting air into the building the other half are letting air out of the building.

The equation is as follows:

$$\left(\frac{\text{Total perimeter lineal feet of crack}}{2}\right) * 0.25 = CFM$$

$$CFM * 1.08 * \Delta T = Btuh$$

That's it.

If you want to calculate the infiltration through an active door, go directly to ASHRAE. There you will find impressive tables with factors for various levels of activity through the door. If you're forced to guess, you might as well guess with the best.

ANECDOTES

Richard's Retrofit Rules

Insulation Rule #1: *"If you burn your hand on the equipment you can probably save some energy."*

Insulation Rule #2: *"If the insulation is in the way of production or maintenance workers it won't last long enough to have a payback."*

Insulation Rule #3: *"If it's already covered with asbestos, the payback period will be measured in decades."*

Insulation Rule #4: *"Sometimes you have to point out the safety features of adding insulation, not the energy saved."*

When I first started doing walk through audits I found myself in plastic molding facilities looking at extruder barrels, dies, and molds that were kept at temperatures from 150° F up to 700° F by electric heat. In none of the first three facilities that I went through did I see insulation on these pieces of equipment.

It didn't take a genius to see, and feel, the potential energy savings. The heat radiating from these parts was awful and in some cases dangerous. Some of the locations were in air conditioned rooms! But being a little cautious, I felt that it would be better to first ask why they weren't insulated. It just seems smarter to first assume that there is a good reason that isn't obvious to an outsider. They may think a question is dumb, but they will think that the *person* is dumb if he makes a recommendation in writing that *they* all know is dumb.

The answers were varied, but mostly consisted of "I don't know." and, "They never are insulated." With some further investigation, it finally came down to two reasons:

1. The extruder barrels only use the electric heaters during the start up of the process run. After the extruder is running up to speed the heat of the friction and the shear forces of the plastic forced through the barrel at high pressures is more than enough to maintain the temperature. In fact, after start up, the barrel has to be cooled with water to keep the temperature from getting too high.

2. They tried insulation many years ago when the only high temperature insulation available was the rigid blocks of extremely brittle material that needed to be custom formed to the equipment. Since the dies, and barrels, etc. require frequent maintenance or production changes, it did not take long for the insulation to be wrecked by excessive handling. At the worst it was a maintenance nightmare. So the practice was abandoned.

It was easy then to recommend the use of custom made blanket type insulation that can withstand high temperatures, is flexible, and is easy to remove and install. Now, I find that many insulation contractors and specialty manufacturers are marketing insulation blankets to these customers.

The only thing that has come to light as a drawback is that many of the older machines were outfitted with electric heater bands that have relatively low temperature electrical insulation on the wires. When these are open to the air, the temperature does not get above their rated design. But when they are covered with insulation, the heat is trapped inside and the electrical insulation can fail.

Chapter 7

Fuel Switching

INTRODUCTION

Fuel switching measures are, for me, some of the most practical approaches to energy management. The real conservation that is desired, for all intents and purposes, is the conservation of money. When you do an evaluation that seeks to identify the equipment and fuel that will accomplish the same results with the least expense you can get creative. In the right situation, with the proper in house support, you can even do some experimenting.

Primarily we are talking about conversion of electric heat to oil or gas. But remember, even though electricity is very expensive, there are times that it is still the best choice. If the utilization is intermittent, if there are great difficulties in getting the necessary pipes and heat exchangers to a remote location, if there are maintenance or contamination problems to worry about, if very high temperatures are required, etc., all of these reasons, and many more, are reason enough to use this expensive source of Btus. But do it on purpose! Use it after careful consideration of all the alternatives. Then you know that you have done your job the right way.

DEFINITIONS

{As you will see, most of these terms are repeated in different sections. I have chosen to repeat them because I felt that each section should be able to stand on its own as a reference. (Not because I needed to fill space in the book.)}

Btu

This stands for British Thermal Unit. A Btu is defined as the amount of heat that it takes to increase the temperature of one pound of water by one degree Fahrenheit. Do not confuse Btu with temperature! The complete combustion of a standard wooden kitchen match will release approximately one Btu. The *temperature* of the flame will be over 1200° F. To get a perspective on quantities of Btus, one gallon of gasoline contains approximately 128,000 Btus.

Btuh

Simply means Btu per hour.

MBH

Simply means 1,000 Btu per hour.

mmBtu and mmBtuh

Means 1,000,000 Btu and Btuh.

Radiation

The emission and propagation of energy through space or through a material medium in the form of waves.

If it's glowing, it's radiating.

Strictly speaking, it doesn't have to be glowing to radiate energy, but in the context of this chapter we are talking about infrared radiation and mostly where the source is hot enough to be radiating some portion of its energy in the visible, red, wavelength.

IR Radiation

Infrared radiation. This is sometimes used to describe a "new" technology in high temperature ovens, etc. The truth is that all ovens with heating elements inside them transfer heat through the radiation of infrared energy. What they are trying to get across is the idea that the equipment has been specifically designed to improve the radiation transfer characteristics of the system.

Another myth is that "Infrared heaters do not heat the air so you automatically save energy." Of course they heat the air. They heat whatever object that the infrared radiation is pointed at and then *that* object heats the air. But how are these heaters controlled? If they are controlled by a thermostat in the room, as most of them are, then the heaters will stay ON until the air in the room is warm enough to satisfy the thermostat setting.

The potential for savings comes from the fact that when you are in a room full of warm objects, the air can be kept much cooler than normal and you will still feel very comfortable.

End-use efficiency

Sources of thermal energy generally have to deal with two areas where heat is lost before it arrives at the "end-use". There are combustion losses, i.e., heat lost up the stack. Then there are transmission losses, i.e., heat lost to the areas that the pipes or ducts pass through before they reach the end-use. The end-use efficiency is used to determine the amount of fuel that needs to be burned at the boiler in order for there to be the required amount of Btus at the end-use. The formula for calculating end-use efficiency is as follows:

*End-Use efficiency = Combustion efficiency * transmission efficiency*

Remember that heat lost by space heating pipes into the space that is to be heated is not really lost. All you lose is control.

Common Conversion Factors:

#6 Fuel Oil	153,000 Btu/gal
#4 Fuel Oil	145,000 Btu/gal
#2 Fuel Oil	138,000 Btu/gal
Gasoline	128,000 Btu/gal
Coal	22,000,000 Btu/ton
Natural Gas	+/- 1,000 Btu/ft^3
Propane	95,000 Btu/gal
Wood	+/- 8,500 Btu/lb.

CALCULATIONS

As stated above, the best situation for cost effective fuel switching projects is when the heat source is electricity: the most expensive source of Btu's known to man. (Actually, the most expensive source of Btu's known to man so far is atomic energy. But we use *that* to make electricity, so by the time we get it, electricity is more expensive.) Electricity has the advantage of being 100% efficient at its end use. Therefore the conversion from units of Btu to units of kWh is very simple:

$$3{,}413\ Btu = 1\ kWh$$
$$\text{or: } 3{,}413\ Btuh = 1\ kW$$

One of the most useful calculations that you can perform allows you to compare all fuel sources by their cost per useful heat content. This is their cost per mmBtu adjusted for their end-use efficiency.

$$\frac{\$ / Unit}{(Btu / Unit) * (End\text{-}use\ Eff.)} * 1{,}000{,}000 = \$ / mmBtu$$

So if electricity costs $0.075/kWh at 100% end-use efficiency and natural gas costs $0.42/CCF (CCF = 100 cubic feet) at 68% end-use efficiency, then the comparison results as follows:

$$\text{Electricity: } \frac{\$0.075}{3{,}413 * 100\%} * 1{,}000{,}000 = \$21.98 / mmBtu$$

$$\text{Natural Gas: } \frac{\$0.42}{(100 * 1{,}000) * 68\%} * 1{,}000{,}000 = \$6.18 / mmBtu$$

Finding the existing peak consumption for the equipment under investigation is usually pretty easy. Most of the time there is a nameplate that tells you the capacity in Btuh, MBH, gallons per hour, cubic feet per hour, or connected electrical load in Watts or kW. With this number known you can size the peak fuel consumption rate required by the replacement equipment. This may be several times higher than the running load for the process, but it is more than likely the rate of heat delivery required for a reasonable start up period.

Therefore, it now becomes necessary to evaluate the running consumption in order to correctly calculate the potential for savings. This is extremely difficult to nail down. If it is an electrically heated device, the use of a recording power meter will give you accurate results for the processes completed over the period of the recording. Whatever the method you use to arrive at an annual consumption value, remember to adjust it to reflect possible differences in end-use efficiency between fuel sources and distribution methods. In many cases when you are comparing similar fuel sources, i.e., natural gas to propane or #4 fuel oil to #2 fuel oil, the annual consumption in Btus will be identical. The operating savings will be the *net* cost of the existing fuel source savings minus the cost to purchase the proposed fuel.

$$\frac{\text{Annual Btu}}{\text{Existing End-use Eff.}} * \text{Existing \$ / Btu} - \frac{\text{Annual Btu}}{\text{Proposed End-use Eff.}} * \text{Proposed \$ / Btu}$$

That's all there is to it. The really hard part is to calculate the cost to install the new equipment. You must take into consideration all the necessary appurtenances to make the conversion complete. A chimney is not usually easy to install as a retrofit item.

Again, this is where creativity can blossom. Remember to let your imagination run past the obvious difficulties. For instance, if they are using electric heat because there is a restriction against an open flame in an explosion proof area, consider installing the burner equipment outside of the restricted area. You can then pipe the heat into the restricted area with some kind of thermal fluid, i.e., steam, hot oil, hot water, etc. Don't forget that you have to include the cost of the energy for the pump when you calculate the operating costs.

ANECDOTES

Richard's Retrofit Rules

Fuel Switching Rule #1: "The production managers think that using oil or gas fuel sources will contaminate the product."

Fuel Switching Rule #2: *"There is no easy way to install a chimney."*

Fuel Switching Rule #3: *"Transporting heat through pipes or ducts is much more difficult, and messy, than using wires."*

Fuel Switching Rule #4: *"'Infrared' is defined by the surface temperature of the infrared source, not the fuel."*

My first fuel switching project involved converting steam boilers that burned #4 fuel oil to "dual fuel" burners. This maintained the #4 fuel burning capability and added the ability to burn natural gas. The natural gas could be purchase *very* cheaply at interruptible rates.

This is actually a very simple project requiring the purchase and installation of packaged components and the extension of a gas line into the building. However, politically I opened a hornet's nest. I found out that my boss had removed dual fuel burners and removed the gas pipe about five years before I was hired. Unfortunately I found out after I had officially made the proposal.

His decision came under the category: "It seemed like a good idea at the time." The burners needed extensive repairs and the gas pipe was in the way of an expansion project. And it was about ten months *before* the first oil embargo. Gas was expensive and oil was *very* cheap.

As you can see, timing is everything. The best answer for a fuel switching project is to leave the existing equipment in place for a back up, if at all possible. Even if it adds some extra cost to the project. It will help production managers to sign on to the project if they think that they can say "I told you so!" and get back to their old ways fast. And you may *need* that back door.

Another project involved fuel switching at the start of the design stage. The company was planning an expansion of both the building and production equipment. The heart of the project was the purchase and installation of a new, larger, faster, high temperature tenter frame (continuous process fabric curing/drying oven).

The current available steam pressure of 100 psi could only produce temperatures of about 340° F. The tenter frame required 400° F. Therefore, the original plans included the typical electrically heated equipment. This would add a connected load of 700 kW which would require a serious increase in the size of the electrical service equipment.

After considerable investigation, it was decided to install a natural gas fired hot oil heater. This would produce oil at 600° F and circulate it through the tenter frame. The additional installation costs were easily offset by the reduction in the electrical service requirements. And the operating savings were significant, even taking into account the operation of the 30 horsepower circulation pumps.

Chapter 8

Heat Recovery

INTRODUCTION

The evaluation of heat recovery conservation measures relies heavily on two things: 1) having a large enough temperature difference between the heat source and the heat sink, and 2) having someplace that you can use the heat if it can be recovered.

This is one of those areas in energy conservation engineering where you can really get your creative juices flowing. Just remember that, while it may have been exciting to be a Pioneer, a lot of Pioneers were killed by Indians. But at this stage of the engineering trail you should be free to explore. The best way to cover your trail is to start off by announcing that you are "only in the brainstorming stage". Then if they start shooting flaming arrows, or just burst out laughing, you can retreat with honor. Or just run.

Heat transfer is a one way street. You can only go from hot to cold. The way you make something hot is to place it next to something hotter. The way you make something cold is to place it next to something colder (that way you make the something colder become hotter and your object, which is hotter, becomes colder – clear?). In all cases, if the two objects are left together long enough they will both become the same temperature somewhere between the two original temperatures. This one of the laws of thermodynamics. I didn't bother to look up which law it is because, honestly, who cares?

It is important to remember that this law exists because it should force you to remember the golden rule: Do Unto Others, Then Leave! In order for there to be a worthwhile project to recover heat (or cold) you must take that energy away immediately. Otherwise your temperature difference driving the heat transfer will eventually disappear. The two things for an ideal situation are a reasonably steady flow of wasted energy *and* a reasonably steady *need* for the recovered energy. This can be overcome to some degree with adequate storage capacity for either side of the transfer. But storage costs capital and loses heat (cold) to standby losses.

High temperature exhaust streams can produce hot water. Process cooling is often accomplished by running city water through a heat exchanger, cooling the process fluid and heating the city water. The more audits you do, the more you'll be surprised to see this clean, hot water dumped down the drain. The first question is whether the plant can use that much hot water. If not, look into a cooling tower and water recirculation system. This will use more energy, but will save major amounts of money in water and sewer charges. (Remember, you are there to *help save money*, not just energy.)

DEFINITIONS

Delta T

No, this is not the name of a Mississippi river boat. The temperature difference between two items or between two locations on the same item is called the Delta T because the Greek letter Δ is used to indicate the mathematical difference between two numbers.

Approach Temperature

This is really the Approach Temperature *Difference*. This is the smallest ΔT between the fluid stream that you are transferring from and the fluid stream that you are transferring to. In essence, it is the closest number of degrees that the target fluid can approach the source fluid to maintain heat transfer. It is a function of the heat transfer characteristics of the heat exchanger under evaluation.

Specifically it allows you to define the lowest (or highest) temperature that the target fluid can achieve. In a cooling tower, it is the ΔT between the ambient air wet bulb temperature and the leaving water temperature. In a heat exchanger it is the expected difference between the entering source fluid temperature and the outlet target fluid temperature.

Specific Heat

This is simply the number of Btus it takes to raise a pound of a substance one degree Fahrenheit. The specific heat of water is 1.0.

THINGS TO LOOK FOR

1. *Stacks above the roof and drainage pits in the floor.* I like to look at a building as an energy conversion box. The energy comes in through pipes and wires and leaves through the building envelope. If it doesn't leave with the product through the loading dock, then it leaves through the walls, out the exhausts, and down the drain.

2. *Water vapor in the air.* Open tanks are very difficult to do anything with, because, in most cases, if they could be closed they would have been closed already. But a fog in the air could also come from a hot water drain line (or steam leaks).

3. *Listen for water splashing or air moving through a duct.* Touch (carefully) the pipes and the ducts. If they are hot or cold find out more.

THINGS TO LOOK OUT FOR

1. *No place to use the recovered heat.* It does no one any good to recover heat from one location only to send down the drain at another location.

2. *Contaminated fluid streams.* Many exhaust streams contain contaminants that are down right scary. There are many reliable approaches to avoiding cross contamination. But remember, no savings are worth being responsible for a health disaster. Secondly, there are contaminants that will stay a

vapor at elevated temperatures. You must be aware of the dew point of these vapors. Some of the contaminants will form corrosive liquids when they condense out of the vapor stage. Others may become pools of flammable liquid that can spontaneously ignite.

CALCULATIONS:

One of the first calculations should be to determine how much energy is available for recovery. The actual formula depends on the medium that is carrying the energy.

Air:

$$Btuh = 1.08 * CFM * \Delta T$$

{The constant, 1.08, is the conversion factor taking the density of air times the specific heat of air times 60 (min./hr). In the real world the density and specific heat will vary depending on the actual conditions. However, this variation is so slight that it will be negligible for all but the most extreme cases.}

Water:

$$Btuh = 500 * GPM * \Delta T$$

{The constant, 500, is the conversion factor taking the density of water (lbs./gal) times the specific heat (1.0) times 60 (min./hr).}

Other liquids (oil, glycol, etc.):

$$Btuh = GPM * (\text{Lbs./Gal.}) * \text{Specific Heat} * \Delta T * 60\ min/hr$$

The questions that you have to answer are specific to the situation: what is the GPM or CFM and what is the ΔT? But as I said before: that's what you get paid for.

ANECDOTES

Richard's Retrofit Rules

Heat Recovery Rule #1: *"The heat source will always be the furthest distance it can be from where you can use it."*

Heat Recovery Rule #2: *"The waste heat stream with the most energy is the one with the most dangerous chemicals."*

Heat Recovery Rule #3: *"If the project goes smoothly during installation the output temperatures will be lower than expected."*

Heat Recovery Rule #4: *"If you find a large source of wasted energy you will not be able to find a matching use for it."*

Heat Recovery Rule #5: *"Production managers do not trust 'second hand' energy."*

My first heat recovery project was in a textile dyehouse. My first month on the job as Corporate Energy Manager I observed clean, hot water going down the drain. Not just a little: 75 gpm at 140°, from three dye machines at a time. This was the cooling water discharge from a shell and tube heat exchanger that cooled the dye bath water from 205° to 100°. This was clean, fresh water.

I then found out about a storage tank, piping, and pumping system that had been installed to store this clean, hot water for use when the process called for hot water. This whole system had been installed, used for a while, then abandoned in place. The reason, as told to me by the dyehouse manager, was that the hot water made the pipes rust more than the cold water. Which made no sense at all to me.

The reason that they believed this was that several dyelots had several yards of fabric ruined by rust, and, of course, "That never happened until we started using the recovered hot water."

Fortunately, the position of Energy Manager was new to this company and I was the first one. This meant that politically I was favored, at least for a while. Therefore, I was able to get them to reactivate the recovery system. This went on for about two weeks until the dyehouse manager came into my office with about ten yards of fabric to show me the rust that was ruining his product.

He unfolded the fabric and exposed a large collection of flakes of rust, both tiny and up to 1/4" across. I picked up the largest flake and remarked that it was incredibly flat to have come from the inside of a pipe. I then turned it over and discovered that one side was painted gray. I then pointed out that I was pretty sure that the inside of the water pipes had not been painted recently. Finally, after he told me which dye machine this had come from, I reminded him that *that* machine was the only machine that was *not* connected to the recovered hot water system.

This is the source of Rule #5. All this evidence must have been there all those earlier times that rust showed up. But their prejudice against second hand heat blinded them.

So where did the rust come from? It took some observation time, and a serendipitous coincidence of being in the right place at the right time, but I finally solved the mystery.

The dye baths involved were open tanks that the fabric sat in and the dye was recirculated through the tank. One day I stopped to ponder the rust question at a tank that was full of clean, standing water and no fabric. I observed tiny flakes of rust *floating* on the surface. I was boggled by the thought that rust introduced through a submerged fill pipe would float to the surface. So I bent over and touched one of them which broke the surface tension of the water and the flake sank to the bottom. Just as I was beginning to realize the truth, a large flake that was about 1-1/2" across went *splook* into the water in front of my face and drifted to the bottom like a leaf in a light breeze.

I looked up and observed the two ton hoist on a monorail above all five tanks for loading and unloading the 2,000 pound (dry weight) dyelots. This monorail was painted gray. I emphasize "*was*" because the effect of the clouds of steam from the dye baths had rusted most of the paint and as a dyelot was moved with the hoist the monorail would flex and more rust and paint would fall into the dye baths.

The solution was to institute an annual preventive maintenance schedule of scraping a repainting the monorail.

A second heat recovery project involved recovering hot water from the exhaust of a tenter frame (an oven for drying fabric and/or curing the chemicals added for various reasons). This oven routinely operated at 400° F with about 12,000 cfm exhaust. By installing a water coil in the exhaust I was able to reclaim close to 80 gpm of 120° water for use in the dyehouse.

All went very well until about three or four weeks passed. On Monday morning, at about 9:00 am, the duct caught fire just down stream from the coil. Because it was contained in the ductwork it wasn't too serious. The fire department put it out reasonably quickly and there was no collateral damage.

In about three days the duct was replaced and the operation was back in service. Until another three or four weeks passed and the duct caught on fire again in the morning.

What was happening? The chemicals from all the different dyes, and fabric treatments, etc. were condensing in the ductwork then cooling down further when the oven was off during third shift. This chemical soup was turning into a witches brew that would spontaneously ignite when it was reheated enough times.

Certainly the potential for heat recovery was greatly increased during the fire, but the city fire department did not appreciate their role in this heat recovery project. They made it clear that they would no longer provide the water supply without some hefty labor charges. And that, of course, made the payback period much too high.

Factory Mutual is *still* scratching their heads over that one.

Richard's Retrofit Rules

LIGHTING

Lighting Rule #1: *"You can never save more energy than shutting it off."*

Lighting Rule #2: *"There is at least one person who won't like it."*

Lighting Rule #3: *"Always retrofit lighting at night."*

Lighting Rule #4: *"The occupancy sensor will turn the lights off when the company president is in the bathroom."*

PUMPING

Pumping Rule #1: *"You can never save any more energy than shutting it off."*

Pumping Rule #2: *"You will never get a production manager to accept downsizing."*

Pumping Rule #3: *"No one in production knows the flow requirements."*

FANS

Fan Rule #1: *"You can never save any more energy than shutting it off. But when you do, someone will complain that it's too ________ (stuffy, cold, hot, quiet, etc.)."*

Fan Rule #2: *"It is always a surprise to find out that the fan noise is more important than the actual movement of air."*

Fan Rule #3: *"Employees always believe that more exhaust is better."*

MOTORS

Motor Rule #1: *"You can never save any more energy than shutting it off."*

Motor Rule #2: *"No one in the plant knows how much the motor is loaded."*

Motor Rule #3: *"If you burn your hand when you touch it, it is at least 100% loaded."*

INSULATION

Insulation Rule #1: *"If you burn your hand on the equipment you can probably save some energy."*

Insulation Rule #2: *"If the insulation is in the way of production or maintenance workers it won't last long enough to have a payback."*

Insulation Rule #3: *"If it's already covered with asbestos, the payback period will be measured in decades."*

Insulation Rule #4: *"Sometimes you have to point out the safety features of adding insulation, not the energy saved."*

FUEL SWITCHING

Fuel Switching Rule #1: *"The production managers think that using oil or gas fuel sources will contaminate the product."*

Fuel Switching Rule #2: *"There is no easy way to install a chimney."*

Fuel Switching Rule #3: *"Transporting heat through pipes or ducts is much more difficult, and messy, than using wires."*

Fuel Switching Rule #4: *"'Infrared' is defined by the surface temperature of the infrared source, not the fuel."*

HEAT RECOVERY

Heat Recovery Rule #1: *"The heat source will always be the furthest distance it can be from where you can use it."*

Heat Recovery Rule #2: *"The waste heat stream with the most energy is the one with the most dangerous chemicals."*

Heat Recovery Rule #3: *"If the project goes smoothly during installation the output temperatures will be lower than expected."*

Heat Recovery Rule #4: *"If you find a large source of wasted energy you will not be able to find a matching use for it."*

Heat Recovery Rule #5: *"Production managers do not trust 'second hand' energy."*

Index